OCTONION-LIKE DNA-BASED LIFE, UNIVERSE EXPANSION IS DECAY, EMERGING NEW PHYSICS

Supplement to *Beyond Octonion Cosmology*

Stephen Blaha Ph. D.
Blaha Research

NEWQUeST Prefigures DNA-based Life
Structure: NEWQUeST ↔ Two Strand, Four Base Pair DNA Fragment
A Detailed Octonion Space Spectrum Analysis
Elaboration on Generation & Layer Groups
Universe Particles: Expansion from the Big Bang
Universe WKB, Vacuum Polarization, and Breit-Wigner Periods
Detailed View of NEWQUeST (Universes)
Detailed View of NEWUTMOST(Megaverses)

Pingree-Hill Publishing
MMXXI

Rev. 00/00/01   June 27, 2021

To Margaret

Some Other Books by Stephen Blaha

All the Megaverse! Starships Exploring the Endless Universes of the Cosmos using the Baryonic Force (Blaha Research, Auburn, NH, 2014)

SuperCivilizations: Civilizations as Superorganisms (McMann-Fisher Publishing, Auburn, NH, 2010)

All the Universe! Faster Than Light Tachyon Quark Starships & Particle Accelerators with the LHC as a Prototype Starship Drive Scientific Edition (Pingree-Hill Publishing, Auburn, NH, 2011).

Unification of God Theory and Unified SuperStandard Model THIRD EDITION (Pingree Hill Publishing, Auburn, NH, 2018).

The Exact QED Calculation of the Fine Structure Constant Implies ALL 4D Universes have the Same Physics/Life Prospects (Pingree Hill Publishing, Auburn, NH, 2019).

Unified SuperStandard Theory and the SuperUniverse Model: The Foundation of Science (Pingree Hill Publishing, Auburn, NH, 2018).

Quaternion Unified SuperStandard Theory (The QUeST) and Megaverse Octonion SuperStandard Theory (MOST) (Pingree Hill Publishing, Auburn, NH, 2020).

Unified SuperStandard Theories for Quaternion Universes & The Octonion Megaverse (Pingree Hill Publishing, Auburn, NH, 2020).

The Essence of Eternity: Quaternion & Octonion SuperStandard Theories (Pingree Hill Publishing, Auburn, NH, 2020).

A Very Conscious Universe (Pingree Hill Publishing, Auburn, NH, 2020).

From Octonion Cosmology to the Unified SuperStandard Theory of Particles (Pingree Hill Publishing, Auburn, NH, 2020).

Beyond Octonion Cosmology (Pingree Hill Publishing, Auburn, NH, 2021).

Available on Amazon.com, bn.com Amazon.co.uk and other international web sites as well as at better bookstores (through Ingram Distributors).

CONTENTS

FIGURES and TABLES

Figure 3.2.. Four layers of Internal Symmetry groups in NEWQUeST (omitting Connection groups). The groups in each layer are independent of those in other layers. The groups in each block of each layer are independent of those in the other blocks.

Introduction

This book is a major supplement to *Beyond Octonion Cosmology*. It provides more detail on important topics in Octonion Cosmology. Octonion Cosmology has been shown to give a complete theory of Elementary Particles and Gravitation for our universe and the enclosing Megaverse. The topics in this book include:

1. The theory of our universe, NEWQUeST, exhibits a form similar to the form of a two strand (double helix) DNA four base pair fragment. NEWQUeST may be said to prefigure DNA-based Life in our universe, and possibly in the Megaverse—modulo the eight dimension space-time of the Megaverse that might influence DNA chemistry. Life is favored at the deepest level!

2. It provides a detailed description of the expansion of the universe as a particle resonance. It shows that in the beginning an essential singularity caused the universe to initially appear as a WKB wave function. In the intermediate expansion period, the universe has the form of a QED-like vacuum polarization. In the recent period, the universe resembles a decaying Breit-Wigner resonance. Thus its particle-like nature.

3. New important features of the Octonion Spaces Spectrum are discussed.

4. Generation, Layer, and Connection group representations are discussed in some detail.

5. Detailed, yet concise, descriptions of NEWQUeST (Universes), and of NEWUTMOST (Megaverses) are presented.

6. The 256 NEWQUeST fundamental fermions are fully interconnected by the Connection groups. Thus no fermions are truly Dark, although some fermions are currently inaccessible due to large masses and/or weak coupling constants.

1. New View of the Octonion Spaces Spectrum

In Blaha (2019c) we created the Octonion Space Spectrum, which is reproduced in Fig. 1.1. In this chapter we investigate the form of the spectrum in more detail focusing on the NEWQUeST space. In particular, we consider the spectrum of octonion spaces from two perspectives:

1. The generation of a space and its instance through fermion-antifermion annihilation in a "higher" space. (The approach of previous books)

2. The strict generation of spaces by Cayley number considerations.

In the absence of experimental data, the definition of the spectrum of octonion spaces is somewhat problematic. Either approach stated above is reasonable. On physical grounds, and due to the importance of the 8th Octonion Space Postulate in *Beyond Octonion Cosmology*:

8. Fermion-antifermion annihilation into space instances supports identifying the form of octonionic subspace dimension arrays (rows and columns) with the form of spinor arrays in the fermion-antifermion annihilation space. In the notation of chapter 4:

$$d_d^{(n)} = d_s^{(n+1)} \qquad (2.1)$$

for the space corresponding to Cayley-Dickson number n. The sizes of the dimension arrays of spaces are directly determined by the Cayley-Dickson numbers. A dimension array is the composition of a hypercomplex Cayley number with itself. Consequently the total number of dimensions d_d of a space, including internal symmetry and space-time dimensions, is $d_d = 2^{2n}$ where n is the Cayley-Dickson number of the space.

the author believes perspective 1 is the correct choice.

1.1 The Octonion Spaces from the view of Perspective 1

Figs. 1.1, 1.2, and 1.3 from *Beyond Octonion Cosmology* display the octonion spectrum based on perspective 1. Section 1.2 considers the octonion spectrum from the view of perspective 2. Based on that discussion we amend the view of perspective 1 in section 1.3 to give space 6 (our universe) six space-time dimensions, which is then reduced to four space-time dimensions by transferring two dimensions to internal symmetries as in chapter 4 of *Pioneering Octonion Cosmology II.*[1]

---

[1] Al;ternately it is possible that two of the six dimensions are compactified,

EIGHT OCTONION SPACES SPECTRUM

Octonion Space Number O_s	Cayley-Dickson Number n	Cayley Number Dimension d_c	Dimension Array d_d	Space-time-Dimension r	Fermion Spinor Array Size d_s	Cayley Number "Name"
0	10	1024	1024×1024	18	512×512	Complex Octonion Octonion Octonion
1	9	512	512×512	16	256×256	Octonion Octonion Octonion
2	8	256	256×256	14	128×128	Quaternion Octonion Octonion
3	7	128	128×128	12	64×64	Complex Octonion Octonion
4	6	64	64×64	10	32×32	Octonion Octonion
5	5	32	32×32	8	16×16	Quaternion Octonion
6	**4**	**16**	**16×16**	**4**	**4×4**	**Complex Octonion**
7	3	8	8×8	4	4×4	Octonion

Figure 1.1. The perspective 1 spectrum of the octonion spaces. The full perspective 1 set of spaces is listed in Fig. 1.2 below. The spaces shown are numbered from 0 through 7. The octonion space number O_s was assigned in our earlier books. The other columns are: Cayley-Dickson number n; the Cayley number dimension d_c for the number of components of the Cayley number; the number of array components d_d of the corresponding array of space dimensions; the space-time dimension r of the space-time; the number of fermion spinor array components d_s for a space-time of dimension r; and the Cayley number name.

TEN OCTONION SPACES SPECTRUM

Spectrum Number

	Coordinate Cayley Type	Dimension of a Coordinates	Dimension Array Size d_d	Space-Time Dimension r
	Superverse Space			
0	Complex Octonion Octonion Octonion (1024)	Complex Octonion Octonion Octonion	1024×1024	18
1	Octonion Octonion Octonion (512)	Octonion Octonion Octonion	512×512	16
2	Quaternion Octonion Octonion (256)	Quaternion Octonion Octonion	256×256	14
3	Complex Octonion Octonion (128)	Complex Octonion Octonion	128×128	12
4	Octonion Octonion (64)	Octonion Octonion	64×64	10
	Maxiverse Space			
5	Quaternion Octonion (32)	Quaternion Octonion	32×32	8
	Megaverses Space			
6	Complex Octonion (16)	Complex Octonion	16×16	6 | 4
	Universe Space			
	Minispaces			
7	Quaternion (4)	Quaternion	4×4	4
8	Real (4)	Real (4)	4×4	4
9	Real (4)	Real (4)	4×4	4
10	Real (4)	Real (4)	4×4	0

Figure 1.2. The spectrum of God-Space (the Superverse) and the ten octonion spaces. The spaces are numbered from 0 through 10. The numbers in parentheses in column 2 are the number of array row/column dimensions. The items in column 3 are the number of rows of dimensions (1024, 512, 256, 128, 64, 32, 16, 4, 4, 4, 4).

God-Space – Space 0: Complex Octonion Octonion Octonion

1,048,576 dimensions $2^{20/2} = 1024$ rows/columns	18 Space-time Dimensions	512 512-spinors

Space 1: Octonion Octonion Octonion

262,144 dimensions $2^{18/2} = 512$ rows/columns	16 Space-time Dimensions	256 256-spinors

Space 2: Quarternion Octonion Octonion

65,536 dimensions $2^{16/2} = 256$ rows/columns	14 Space-time Dimensions	128 128-spinors

Space 3: Complex Octonion Octonion

16,384 dimensions $2^{14/2} = 128$ rows/columns	12 Space-time Dimensions	64 64-spinors

Space 4: Octonion Octonion

4,096 dimensions $2^{12/2} = 64$ rows/columns	10 Space-time Dimensions	32 32-spinors

Space 5: Quaternion Octonion

1,024 dimensions $2^{10/2} = 32$ rows/columns	8 Space-time Dimensions	16 16-spinors

Space 6: Complex Octonion

256 dimensions $2^{8/2} = 16$ rows/columns	4 Space-time Dimensions	4 4-spinors

Space 7: Octonion

64 dimensions $2^{6/2} = 8$ rows/columns	2 Space-time Dimensions	4 4-spinors (Built from four 16 dimension spaces of the 10 space spectrum)

Figure 1.3.The sequence of fermion-antifermion annihilations and the set of spaces that they generate. The eleven space sequence would have a modified

space 7 and add three more spaces, 8, 9, and 10, to the sequence as shown in Fig. 1.2.

1.2 The Octonion Spaces from the view of Perspective 2

In section 1.1 we arbitrarily set the Cayley number n = 4 (NEWQUeST) space-time dimension to four, the dimension of our universe. See Fig. 1.1. We did *not* use the rule relating Cayley number n to the number of space-time dimensions for n = 4:

$$r = 2n - 2 \qquad\qquad (1.1)$$

We now use eq. 1.1 throughout the octonion spectrum of spaces. In addition we also assume the n = 4 space (NEWQUeST) emerges with four dimensions in conformity with our universe due to the transfer of dimensions to internal symmetry. The result is Figs. 1.4 and1.5. Fig. 1.5 is based on perspective 2.

EIGHT OCTONION SPACES SPECTRUM

Octonion Space Number o_s	Cayley-Dickson Number n	Cayley Number Dimension d_c	Dimension Array d_d	Space-time-Dimension r	Fermion Spinor Array Size d_s	Cayley Number "Name"
0	10	1024	1024×1024	18	512×512	Complex Octonion Octonion Octonion
1	9	512	512×512	16	256×256	Octonion Octonion Octonion
2	8	256	256×256	14	128×128	Quaternion Octonion Octonion
3	7	128	128×128	12	64×64	Complex Octonion Octonion
4	6	64	64×64	10	32×32	Octonion Octonion
5	5	32	32×32	8	16×16	Quaternion Octonion
6	4	16	16×16	$6 \rightarrow 4$	$8 \times 8 \rightarrow 4 \times 4$	**Complex Octonion**
7	3	8	8×8	4	4×4	Octonion

Figure 1.4. The slightly revised spectrum of the octonion spaces using eq. 1.1. The octonion space number o_s was assigned in our earlier books. The other columns are: Cayley-Dickson number n; the Cayley number dimension d_c for the number of components of the Cayley number; the number of array components d_d of the corresponding array of space dimensions; the space-time dimension r of the space-time; the number of fermion spinor array components d_s for a space-time of dimension r; and the Cayley number name.

THE TEN OCTONION SPACES SPECTRUM

Octonion Space Number O_s	Cayley-Dickson Number n	Cayley Number Dimension d_c	Dimension Array d_d	Space-time-Dimension r	Fermion Spinor Array Size d_s	Cayley Number "Name"
0	10	1024	1024×1024	18	512×512	Complex Octonion Octonion Octonion
1	9	512	512×512	16	256×256	Octonion Octonion Octonion
2	8	256	256×256	14	128×128	Quaternion Octonion Octonion
3	7	128	128×128	12	64×64	Complex Octonion Octonion
4	6	64	64×64	10	32×32	Octonion Octonion
5	5	32	32×32	8	16×16	Quaternion Octonion
6	**4**	**16**	**16×16**	$6 \rightarrow 4$	$8 \times 8 \rightarrow 4 \times 4$	**Complex Octonion**
7	3	8	8×8	4	4×4	Octonion
8	2	4	4×4	2	2×2	Quaternion
9	1	2	2×2	0	1×1	Complex
10	0	1	1×1	$-2 \rightarrow 0?$	1×1	Real

Figure 1.5. A ten space spectrum based on perspective 2.

Space 10 in Fig. 1.5 specifies a space-time dimension of -2, which is unphysical from current viewpoints. If we now invoke perspective 1 basing the spectrum on fermion-antifermion space instance creation, then we obtain Fig. 1.6.

TEN OCTONION SPACES SPECTRUM
(based on the Pattern of Fermion-Antifermion Annihilation)

Spectrum Number

	Coordinate Cayley Type	Dimension of a row/column	Dimension Array Size d_d	Space-Time Dimension r
0	Complex Octonion Octonion Octonion (1024)	Complex Octonion Octonion Octonion	1024×1024	18
1	Octonion Octonion Octonion (512)	Octonion Octonion Octonion	512×512	16
2	Quaternion Octonion Octonion (256)	Quaternion Octonion Octonion	256×256	14
3	Complex Octonion Octonion (128)	Complex Octonion Octonion	128×128	12
4	Octonion Octonion (64) Maxiverse Space	Octonion Octonion	64×64	10
5	Quaternion Octonion (32) Megaverses Space	Quaternion Octonion	32×32	8
6	Complex Octonion (16) Universe Space **Minispaces**	Complex Octonion	16×16	$6 \rightarrow 4$
7	Quaternion (4)	Quaternion	4×4	4
8	Real (4)	Real (4)	4×4	4
9	Real (4)	Real (4)	4×4	4
10	Real (4)	Real (4)	4×4	0

Figure 1.6. The 11 spaces spectrum of God-Space – space 0 (also called the Superverse) and the ten octonion spaces. The spaces are numbered from 0

through 10. The numbers in parentheses in column 2 are the array dimensions. The items in column 3 are the number of rows/columns. The derivation and interpretation of spaces 7, 8, 9, and 10 are provided in *Beyond Octonion Cosmology*. They are changed from Fig. 1.5 above. Space 10 has zero space-time dimensions to furnish a cutoff of space instance generation.

Fig. 1.6 constitutes the octonion spaces spectrum combining perspectives 1 and 2 as the suthor views it.

2. Universe Expansion as a Particle Decay

2.1 General Picture of the Annihilation Process

The transition from a fermion-antifermion pair in a higher space to an octonion space instance in its lower space, and its associated space definition is a worthy subject for deeper investigation. Quantum Field Theory provides a description of the process that appears to be (initially) satisfactory for physical purposes. But the description of the process is but a somewhat superficial description. The reality of the transformation, and the reality of all particle creation, transformation, and annihilation processes, does not appear to be fully captured by the Quantum Field Theory formalism.

Processes, viewed from a quantum field theory perspective, *seem* to be described satisfactorily because they take place at the most infinitesimal level and thereby escape infinitesimal "details" that would further elucidate the evolution of creation and annihilation processes.

Thus we have only a rudimentary, although physically satisfactory for the most part, understanding of creation and annihilation. We are up against the wall of ignorance that Reality gives.

In considering this wall of ignorance we must realize:

1. No theory is exact.

2. There is an unstated, but understood, "uncertainty" relation that we face

$$\Delta \text{Observation } \Delta \text{Reality} \geq \text{“}\hbar\text{”}$$

where $\hbar$ is in quotes since it can only be viewed as a symbolic use of the Heisenberg constant.

This uncertainty relation tells us the "spread" of observations is related to the "spread" of our understanding of Reality. The greater the spread (range) of observations, the more precise our understanding of the reality of an event.

With this basis of understanding, we turn to examine the process of annihilation of fermions into a space instance. The process of the combination of spinors is important since it determines the dimensions of the space of the created instance. The vacuum-polarization-like transition of the created instance to an expanding universe is also of great interest.

2.2 Combination of Fermion and Antifermion Spinors

The spinors of a colliding fermion – antifermion pair start with spinors of the type u and v. Their combination was considered in detail in *Beyond Octonion*

Cosmology as well as earlier books by the author. Remarkably their structure carries over to the form of the dimension array of a created universe. Symmetries are divided into 64×64 blocks that are further subdivided into 16×16 blocks.

2.3 Transform of Universe Scale Factor a(t) to a Vacuum Polarization-like Form

In *Universes are Particles* we showed that the expansion of our universe appears to be describable by the Fourier transform of the vacuum polarization of a scalar particle. This fact, and other facts, led us to conclude that a universe is a type of particle that starts as an ultra microscopic particle and then expands as we have seen in our universe.

Octonion Cosmology views universes as the result of fermion-antifermion annihilation in a Megaverse – a higher space in the octonion space spectrum.

In this section we will consider our phenomenological universe scale factor:[2]

$$a(t) = [(t + t_0)/t_{now}]^{g\,d/(t + t_0)] + g + h(t + t_0)} \tag{2.1}$$

where t is the time, $-t_0$ is the instant of universe creation, and g, d, and h are constants fixed by Hubble Parameter data and a Big Bang model[3] of the author.

We will Fourier transform a(t) to find an equivalent "momentum space" form and analyze it as a vacuum polarization.

The constants in a(t) that are fixed by Hubble Parameter data are:

$$h = 2.25983 \times 10^{-18} \tag{2.2}$$
$$g = 0.000282377 = 2.82377 \times 10^{-4}$$

and by our Big Bang model:[4]

$$t_0 = 1.936 \times 10^{-197} \text{ s} \tag{2.3}$$
$$d = 2.956 \times 10^{-194} \text{ s}$$

The scale factor a(t) can be factored into three parts that govern the ultra small time Big Bang period, the intermediate period, and the recent period:

$$a(t) = \underbrace{[(t + t_0)/t_{now}]^{gd/(t + t_0)}}_{\textbf{I}} \underbrace{[(t + t_0)/t_{now}]^{g}}_{\textbf{II}} \underbrace{[(t + t_0)/t_{now}]^{h(t + t_0)}}_{\textbf{III}} \tag{2.4}$$

where the labels I, II, and II indicate the role of each factor in each period of the time evolution of the universe: I indicates the Big Bang period, II indicates an intermediate

---

[2] There are Standard Models of Cosmology models that provide estimates of the Hubble Parameter and a(t). Because of the considerations in *Universes are Particles* that lead to the general form of our a(t), we anticipate these models yield similar features.
[3] Blaha(2004) and (2019e).
[4] Blaha (2021b).

period, and III indicates the "recent" period. The constant t_{now} is the present time, approximately 13.8 Gyears (4.36×10^{17} s.)

2.3.1 The Big Bang Period

This period begins with an essential singularity that serves multiple purposes:

1. It causes $a(t) \to 0$ at $t \to -t_0$ as expected.

2. At an essential singularity all derivatives of $a(t)$ also are zero guaranteeing that nothing carries "forward" from prior to the essential singularity. Thus there is no relevant "before time."

3. The Hubble parameter and all its derivatives becomes infinite at $t = -t_0$. Thus universe expansion may be said to "jump start."

Near the essential singularity the II and III factors of $a(t)$ are slowly varying so we may take them to be approximately constant. The essential singularity factor is rapidly varying:

$$a_I(t) = [(t + t_0)/t_{now}]^{gd/(t + t_0)} \qquad (2.5)$$

We now Fourier transform $a_I(t)$ to find the form of the essential singularity in momentum space. We base the transformation on the energy–time uncertainty relation:

$$\Delta E \Delta t \geq \hbar \qquad (2.6)$$

where we let $E = p^2$. The Fourier transform is

$$a_I(p) = (2\pi)^{\frac{1}{2}} \int_{-t_0}^{\infty} dt/(t + t_0) \ e^{iEt} \ [(t + t_0)/t_{now}]^{gd/(t + t_0)} \qquad (2.7)$$

$$= (2\pi)^{\frac{1}{2}} e^{-iEt_0} \int_{0}^{\infty} dx/x \ e^{iEx} \ [x/t_{now}]^{gd/x} \qquad (2.8)$$

using $x = t + t_0$. The integral in eq. 2.8 is transcendental. To obtain a tractable result we use the inequality

$$[x/t_{now}]^{gd/x} < e^{gd/x} \qquad (2.9)$$

for small x to find an upper bound on $a_I(p)$

.

$$a_I(p) \ < \ (2\pi)^{\frac{1}{2}} \, e^{-iEt_0} \int_0^\infty dx/x \ e^{iEx} \, e^{gd/x} \tag{2.10}$$

We find[5]

$$a_I(p) \ < \ I = (8\pi)^{\frac{1}{2}} \, e^{-iEt_0} \, K_0((igdE)^{\frac{1}{2}}) \tag{2.11}$$

where K_0 is a Bessel function. Approximating K_0 for large E we find

$$
\begin{aligned}
I &= (8\pi)^{\frac{1}{2}} \, e^{-iEt_0} \, i^{1/4} \pi^{\frac{1}{2}}/(2^{\frac{1}{2}}(gdE)^{1/4}) \ \exp[-i^{\frac{1}{2}}(gdE)^{\frac{1}{2}} - i\pi/4] \\
&= 2\pi(i/gdE)^{1/4} \exp[-i(gdE/i)^{\frac{1}{2}} - i\pi/4 - iEt_0]
\end{aligned}
\tag{2.12}
$$

or letting $E = p^2$ we find the absolute value of $a(p)$ is bounded by

$$|a_I(p)| \ < \ |2\pi(i/gd)^{1/4}/p^{\frac{1}{2}} \exp[-i^{\frac{1}{2}}(gd)^{\frac{1}{2}}p - i\pi/4 - ip^2 t_0]| \tag{2.13}$$

It has the form of a WKB wave function in a potential well:

$$\psi = 2p^{-\frac{1}{2}} \sin(\hbar^{-1} \int pdx + \pi/4) \tag{2.14}$$

Comparing to eq. 2.13 we

$$\hbar^{-1} \int pdx \sim (gd)^{\frac{1}{2}}p \equiv \hbar^{-1} \int_0^{gd} Edt \tag{2.15}$$

for constant E implying that gd is the "width" *in time* Δt of the "potential well."

$$\Delta t = gd = 8.347 \times 10^{-198} \ s \tag{2.16}$$

gd which is approximately equal to

$$t_0 = 1.936 \times 10^{-197} \tag{2.17}$$

confirming our view of the interpretation of eq. 2.13 as a WKB approximation:

$$t_0 \cong gd \tag{2.18}$$

Thus we find that $a_I(p)$, in the essential singularity region, is approximately bounded by a WKB wave function for a quasifree particle in a potential.

---

[5] Gradshteyn, I. S. and Ryzhik, I. M., (1965) p.340, 3.471.10 and 3.471.11.

2.3.2 The Intermediate Period

The intermediate period may be defined as

$$t_0 \ll t \ll t_1 \tag{2.19}$$

where t_1 is the estimated point in time of the universe transition from radiation-dominated to matter-dominated:

$$t_1 = 1.48 \times 10^{12} \ s$$

In the intermediate period the first and third factors of a(t) are slowly changing. The second factor dominates the dynamics:

$$a_{II}(t) = [(t + t_0)/t_{now}]^g \tag{2.20}$$

The Fourier transform is

$$a_{II}(p) = (2\pi)^{\frac{1}{2}} \int_0^\infty dt/t \ e^{iEt} \ [t/t_{now}]^g \tag{2.21}$$

$$= (2\pi)^{\frac{1}{2}} k \ [p^2/\Lambda^2]^{-g}$$

where $E = p^2$ and

$$k = \Gamma(2g) \ e^{i\pi g} \tag{2.22}$$

and

$$\Lambda^2 = 1/t_{now} \tag{2.23}$$

with g = 0.000282377. The value of g_{QEDZ2} for Z_2 (the wave function renormalization constant) in the author's exact calculation of the QED Fine Structure Constant α is[6]

$$g_{QEDZ2} = -g_{QEDZ3} = 0.0005805369 \tag{2.23}$$

Thus

$$g = 2 \ g_{QEDZ2} \tag{2.24}$$

If we calculate the value of the exponent in the vacuum polarization of a scalar universe particle,[7] which we denote as g_U as we did in Blaha (2021d), we find

$$g = g_U \tag{2.25}$$

to within 3% where g appears in eq. 2.20 above. *Thus $a_{II}(p)$ equals the vacuum polarization of a scalar particle up to a constant.* Given the approximate nature of our

---

[6] Blaha (2019f) and S. Blaha, Phys. Rev. D 9, 2346 (1974) in particular eq. 65.
[7] Since we assume the universe is a scalar particle: no universal rotation has been detected.

JBW-like calculations of vacuum polarization the agreement is remarkable.[8] *Again support for a particle view of universes!*

2.3.3 The Recent Period

The recent period of universe expansion

$$t_1 < t < t_{now}$$

is dominated by

$$a_{III}(t) = [(t + t_0)/t_{now}]^{h(t + t_0)} \tag{2.26}$$
$$\cong [t/t_{now}]^{ht}$$

since the other factors are slowly varying. The Fourier transform of $a_{III}(t)$ is

$$a_{III}(p) = (2\pi)^{\frac{1}{2}} \int_0^\infty dt/t \ e^{iEt} \ [t/t_{now}]^{ht} \tag{2.27}$$

Since for small t

$$[t/t_{now}] > [t_1/t_{now}] \tag{2.28}$$

in absolute value, and since the large t contribution to the integral is damped by the exponential term, we can approximate eq. 2.28 with the lower bound

$$J = (2\pi)^{\frac{1}{2}} \int_0^\infty dt/t \ e^{iEt} \ [t_1/t_{now}]^{ht} \ < a_{III}(p) \tag{2.29}$$

We find

$$J = i/(p^2 + i \ln(t_{now} /t_1)) \tag{2.30}$$

with $E = p^2$.

J has the form of a relativistic Breit-Wigner distribution for an unstable particle with a propagator denominator

$$1/(p^2 + i \ \Gamma) \tag{2.31}$$

where

$$\Gamma = \ln(t_1 \Lambda^2) \tag{2.32}$$

---

[8] And may be exact! The value of the Hubble Constant H in recent times varies from about 70 – 75 making the calculation of g also approximate. We chose an average value of 73.24 to obtain the value of g above. If we chose the current value for H to be 75.58 we would have $g = -2g_U$ exactly (eq. 25.28). Note: studies of binary black hole merger gravity waves have given a Hubble Constant of 75.2 km s^{-1} Mpc^{-1} (and earlier of 78 km s^{-1} Mpc^{-1}), and studies of light bent by distant galaxies give H = 72.5 km s^{-1} Mpc^{-1}. Thus the value H = 75.58 is not unreasonable.

Thus the universe resembles a decaying resonance particle in recent times.

2.4 The Particle Universe Emerges
The preceding sections demonstrate that the universe resembles a particle. In the earliest period it has a WKB-like particle wave function. In the intermediate period it displays vacuum polarization-like features. In the recent period it resembles a decaying Breit-Wigner resonance.

2.5 Basis of Phenomenological Scale Factor
The motivation for the phenomenological scale factor a(t) is:

1. The known general shape of H(t) at early times, and at present: a *massive decline from the Big Bang period followed by the recent rise.*
2. The simplicity of the model. Two values of H(t) set the constants g and h.
3. Faster than exponential future growth as $t \rightarrow \infty$.

$$a(t) = \exp[(g + ht)\ln(t/t_{now})] \sim e^{ht \ln(t)}$$

4. Power law behavior (in part) as in the radiation dominated approximations for a(t).
5. *The small time behavior of a(t) can be derived in a particle model of a universe as shown in Blaha (2019c).*
6. The determination of the universe vacuum polarization phase of universe growth.
7. The Adler[9] conjecture of an essential singularity in the universe vacuum polarization, which can be used to determine the Big Bang beginning.

Universes and Megaverses grow from annihilation events in a Megaverse and a Maxiverse respectively. The phenomenological scale factor a(t) embodies the growth as far as it is known.

2.6 Conclusion
The universe is like a particle resonance. Universe expansion is a form of resonance decay. The phases of a(t) based expansion are:

$$a(t) = [(t + t_0)/t_{now}]^{gd/(t + t_0)} \cdot [(t + t_0)/t_{now}]^{g} \cdots [(t + t_0)/t_{now}]^{h(t + t_0)}$$

I	II	III
Earliest Time	Intermediate Time	Recent time
Quasi-free particle state	Vacuum polarization state	Decaying Resonance State
Ultra high energy	High energy	low energy

---

[9] S. Adler, Phys. Rev. **D5**, 3021 (1972).

3. NEWQUeST and NEWUST

3.0 Brief Description

Originally the author developed a theory of elementary particles called the Unified SuperStandard Theory (UST). Subsequently the author found that an octonion-based theory called QUeST yielded UST except for a few details. Recently the author modified[10] UST and QUeST by "tradimg" internal symmetry coordinates for space-time coordinates. The results were NEWUST and NEWQUeST.

3.1 Detailed Summary of UST, QUeST, NEWUST, and NEWQUeST

This section summarizes features of these four equivalent theories. The theories can be found in

Blaha, 2018e, *Unification of God Theory and Unified SuperStandard Model THIRD EDITION* (Pingree Hill Publishing, Auburn, NH, 2018).

Blaha, 2020a, *Quaternion Unified SuperStandard Theory (The QUeST) and Megaverse Octonion SuperStandard Theory (MOST)* (Pingree Hill Publishing, Auburn, NH, 2020).

Blaha, 2020c, *Unified SuperStandard Theories for Quaternion Universes & The Octonion Megaverse* (Pingree Hill Publishing, Auburn, NH, 2020).

Blaha, 2021c, *Beyond Octonion Cosmology* (Pingree Hill Publishing, Auburn, NH, 2021).

Blaha, 2021d, *Universes are Particles* (Pingree Hill Publishing, Auburn, NH, 2021).

as well as earlier books by the author.

The theories[11] originated in the past twenty years from the Standard Model of Particles with $SU(2) \otimes U(1) \otimes SU(3)$, and Two-Tier Quantum and PseudoQuantum Field Theory. Noting the presence of conserved particle numbers, and the presence of at least three fermion generations, we introduced U(4) Generation Groups and U(4) Layer Groups together with four layers of four generation fermions. The "Normal" fermions had a matching set of four layers of four generations of "Dark" fermions. The result was the Unified SuperStandard Theory (UST) symmetry:

$$\{[SU(2) \otimes U(1) \otimes SU(3)]^2 \otimes U(4)^4\}^4$$

---

[10] Blaha (2021b) and (2021c).

[11] The author is *solely* responsible for these theories. He has not received any feedback, suggestions, or comments. Contrary to some rumors the author has had no academic affiliation in the past twenty years, and only an honorary affiliation in the preceding 20 years. In the past 10 years the author has not had contact with other physicists. In the past 20 years there were a few very brief contacts of a merely social nature so that the author could pursue his original line of research unencumbered by preconceptions.

with an additional Strong Interaction U(1) group (analogous to that of ElectroWeak theory) found to be needed. Space-time had four dimensions.

In January, 2020 the author discovered that an octonion-based theory that he constructed and named QUeST had the same internal symmetries as UST with the addition of $U(1)^8$. QUeST's internal symmetry, which could be based on a 16×16 dimension array, was

$$[SU(2) \otimes U(1) \otimes SU(3) \otimes U(1)]^8 \otimes U(4)^{16}$$

The addition of $U(1)^8$ indicates that the Strong Interactions in the theory are broken Strong U(4).

During 2020 the author developed an octonion space spectrum that appears in Chapter 1 for both universes and Megaverses, and other spaces. The spectrum was shown to arise from a generation mechanism whereby fermion-antifermion annihilation in a higher space produced an instance of a lower space. A critical part of the derivation of the octonion spectrum was the realization that even space-time dimension spinor arrays are composed of Cayley number rows and columns. Spinor arrays of annihilating fermion-antifermion pairs were shown to generate the dimension arrays of subspace instances. Analyzing the spinor arrays the author noted that the dimension array could be viewed as composed of 64 dimension blocks, which were further subdivided into 16 dimension subblocks.

The subblock structuring, using the known contents of the Standard Model plus Generation and Layer groups, gives the dimension array structure containing 4×4 subblocks in Figs. 3.1 and 3.2.

Thus there was a most satisfactory match between UST and QUeST with the only significant difference being the space-time: four octonion (complex quaternion) coordinates for QUeST and four real space-time coordinates for UST.

Since we see only real dimensions in Reality, we recently transferred 28 QUeST dimensions from space-time to $U(2)^7$ internal symmetry dimensions. The set of internal symmetries was increased by $U(2)^7$, which we call Connection Groups. Each Connection group specifies interactions between corresponding fermions (e with e, q with q, and so on) in separate layers and between Normal and Dark fermions. The connections between the various blocks of fermions is shown in Fig. 3.3. We implement the very practical rule that all blocks must be connected by interactions or they would not be of physical interest. A totally isolated block effectively does not exist physically (except possibly for gravitation effects).

The interactions of the Connection groups must be very weak and/or their gauge bosons must be very massive.

The addition of the Connection Groups and the reduction of space-time dimensions accordingly results in NEWQUeST and NEWUST as summarized below. See appendix 3-A for a discussion of NEWQUeST groups.

Note: the Generaation, Layer, and Connection groups are all badly broken. Their vector bosons must be very massive since they have not been detected in experiments.

3.1.1 Internal Symmetries

The groups are ElectroWeak SU(2)⊗U(1), Strong SU(3), Generation Group U(4), Layer Group U(4), and U(2) and U(4) Connection groups obtained by transfer from space-time coordinates (See Blaha 2012c). The SU(3)⊗U(1) symmetry may be a broken SU(4) symmetry in eqs. 3.1 – 3.4. The internal symmetries for the theories are:

UST

$$[SU(2) \otimes U(1) \otimes SU(3)]^8 \otimes U(4)^{16} \tag{3.1}$$

QUeST

$$[SU(2) \otimes U(1) \otimes SU(3) \otimes U(1)]^8 \otimes U(4)^{16} \tag{3.2}$$

NEWQUeST

$$[SU(2) \otimes U(1) \otimes SU(3) \otimes U(1)]^8 \otimes U(4)^{16} \otimes U(2)^7 \tag{3.3}$$

The only change is in internal symmetries: Twenty-eight real dimensions transferred from space-time coordinates to Connection group $U(2)^7$ internal symmetry.

NEWUST

$$[SU(2) \otimes U(1) \otimes SU(3) \otimes U(1)]^8 \otimes U(4)^{16} \otimes U(2)^7 \tag{3.4}$$

The only change in internal symmetries: Twenty-eight real dimensions added for $U(2)^7$ Connection group internal symmetry.

3.1.2 Space-Time Coordinates

UST

Four real space-time coordinates.

QUeST

Four octonion (complex quaternion) coordinates.

NEWUST

Four real space-time coordinates. No change from UST space-time.

NEWQUeST

Four real space-time coordinates. The six coordinates in the n = 4 octonion space (Figs. 1.5 and 1.6) were lowered to four space-time coordinates with two coordinates transferred to Connection groups.

The only change is in space-time coordinates: Fourteen dimensions transferred from QUeST space-time coordinates to Connection group $U(2)^7$ internal symmetry.

3.2 Fundamental Fermion Spectrum

There are 256 fundamental fermions in NEWQUeST and NEWUST. Conceptually their structure can be viewed as an extrapolation of the known three generations of The Standard Model. For reasons given previously and in Appendix 3-A a fourth generation was indicated and a corresponding Dark sector of similar structure was added. In addition because of the need for Layer groups (Appendix 3-A) the overall structure consisted of four copies of this layer.

Correspondingly, each layer also has its own set of internal symmetry gauge groups to limit mixing between the layers to Layer group interactions and Connection group interactions.

Fig. 3.4 shows the structure of the NEWQUeST/NEWUST fermions. The blocks are 4×4 reflecting the origin of the NEWQUeST/NEWUST space (universe) instance from Megaverse fermion-antifermion annihilation. The spinor analysis of their spinor arrays yields a 16 dimension-based block structure. The 64 dimension fermion layers reflects the 64 dimension structuring of the Megaverse obtained from its creation by fermion-antifermion creation in the Maxiverse. . .

3.3 Total Dimensions

The total of internal symmetry and space-time dimensions is 256 in all four theories listed above.. It is based on the 16×16 dimension array of the Cayley number $n = 4$ space of the Octonion spectrum.

3.4 Pattern of Internal Symmetries

The NEWQUeST dimension array is subdivided into four layers of 56 dimensions—just as in NEWUST (and UST). Fig. 3.1 displays the layers using SU(4) in place of SU(3)⊗U(1). Each layer has a block of 28 dimensions for Normal and 28 dimensions for Dark sectors. There are also seven U(2) Connection groups[12] plus four real-valued space-time coordinates bringing the NEWQUeST total to 256 dimensions.= 4*56 + 28 + 4 = 256 dimensions. The Connection groups are shown in Fig. 3.2, which is a revision of the pattern shown in Blaha (2021c).

---

[12] See Appendix 3-A.

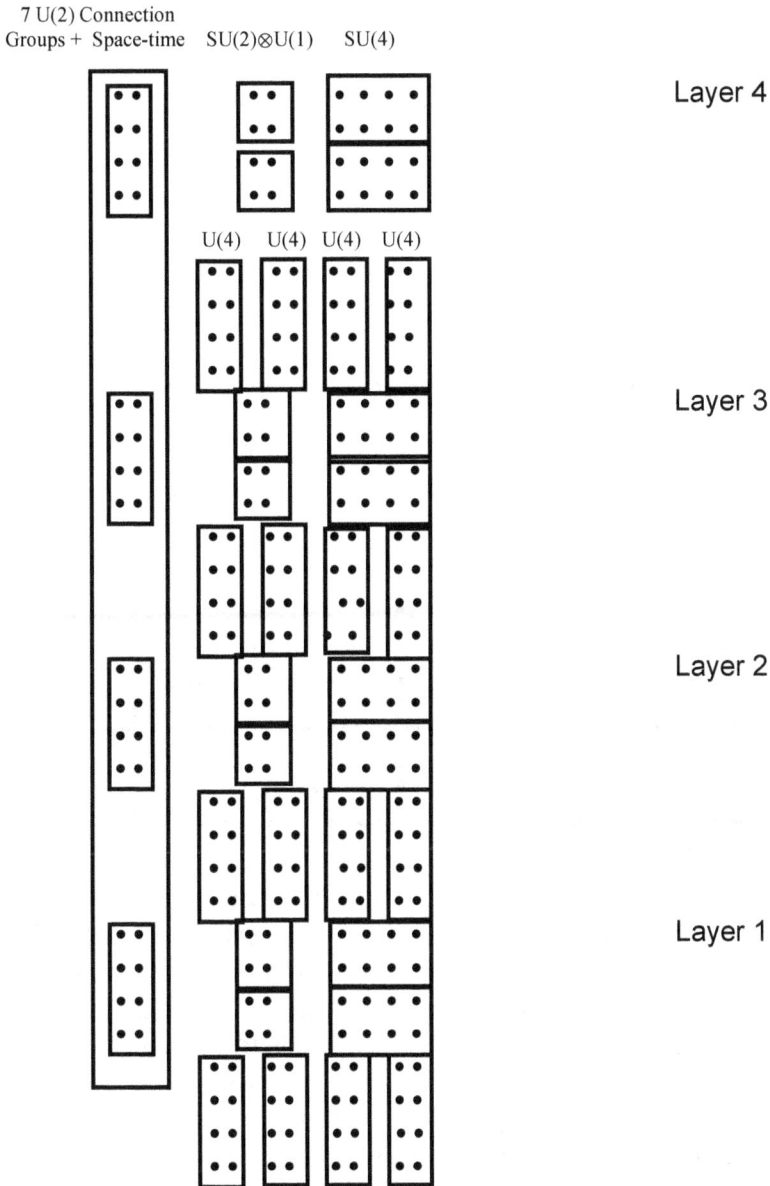

Figure 3.1. The four layers of NEWUST and NEWQUeST internal symmetry groups (and space-time) with SU(4) before breakdown to SU(3)⊗U(1). Note the left column of blocks combine to specify a 4 dimension real space-time plus seven Connection groups. Note each layer has 64 dimensions = 56 + 8 dimensions.

Layers	NORMAL		DARK	
	4	4	4	4
4	SU(2)⊗U(1)⊗SU(3)⊗U(1) 4 Space-time Dimensions	Generation + Layer Groups	SU(2)⊗U(1)⊗SU(3)⊗U(1) 4 Space-time Dimensions	Generation + Layer Groups
4	SU(2)⊗U(1)⊗SU(3)⊗U(1) 4 Space-time Dimensions	Generation + Layer Groups	SU(2)⊗U(1)⊗SU(3)⊗U(1) 4 Space-time Dimensions	Generation + Layer Groups
4	SU(2)⊗U(1)⊗SU(3)⊗U(1) 4 Space-time Dimensions	Generation + Layer Groups	SU(2)⊗U(1)⊗SU(3)⊗U(1) 4 Space-time Dimensions	Generation + Layer Groups
4	SU(2)⊗U(1)⊗SU(3)⊗U(1) 4 Space-time Dimensions	Generation + Layer Groups	SU(2)⊗U(1)⊗SU(3)⊗U(1) 4 Space-time Dimensions	Generation + Layer Groups

Figure 3.2.. Four layers of Internal Symmetry groups in NEWQUeST (omitting Connection groups). The groups in each layer are independent of those in other layers. The groups in each block of each layer are independent of those in the other blocks. Each block contains 16 dimensions. The block dimensions furnish fundamental representations for the groups listed. The entire set of blocks contains 256 dimensions. Each layer contains 56 internal symmetry dimensions. The first two columns are for the "Normal" sector. The last two columns are for the "Dark" sector (although most of the Normal sector is Dark observationally at present.) This Figure also holds for UST with the addition of U(1) groups. The eight sets of 4 real dimension space-times combine to give a 4 real dimension space-time and seven U(2) Connection groups.

Connection Group Applied to Fermions

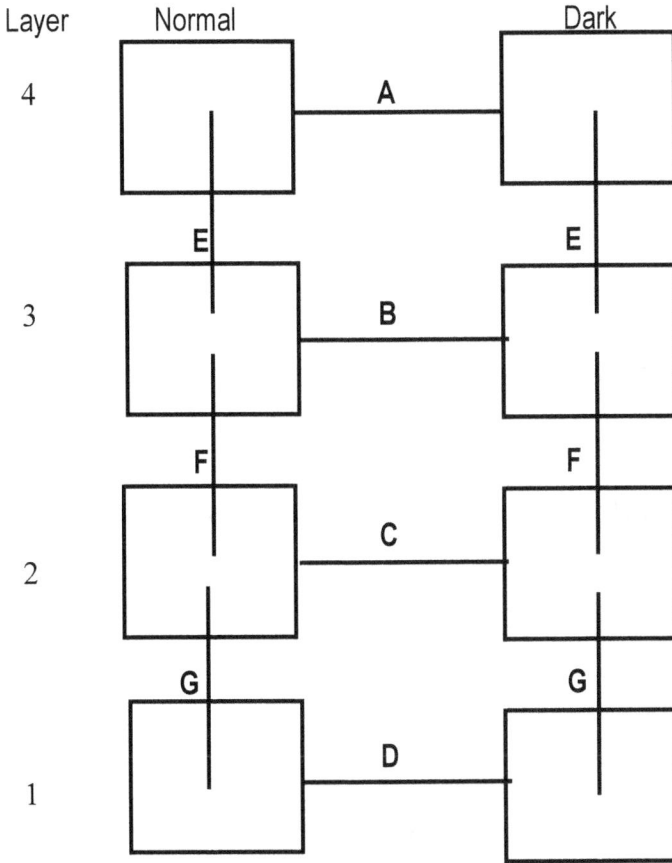

Figure 3.3. The seven U(2) Connection groups (shown in 10 lines) between the eight NEWQUeST/NEWUST blocks. Connection groups are obtained by transfering 28 dimensions from QUeST space-time to internal symmetries with the consequent reduction of the space-time from four octonion (complex quaternion) coordinates to four real coordinates. The Connection groups generate rotations and interactions between corresponding fermions and vector bosons of each pair of blocks. See Appendix 3-A. The Normal and Dark sector U(2) vertical connections above (E, F, G) represent the same U(2) group layer by layer.

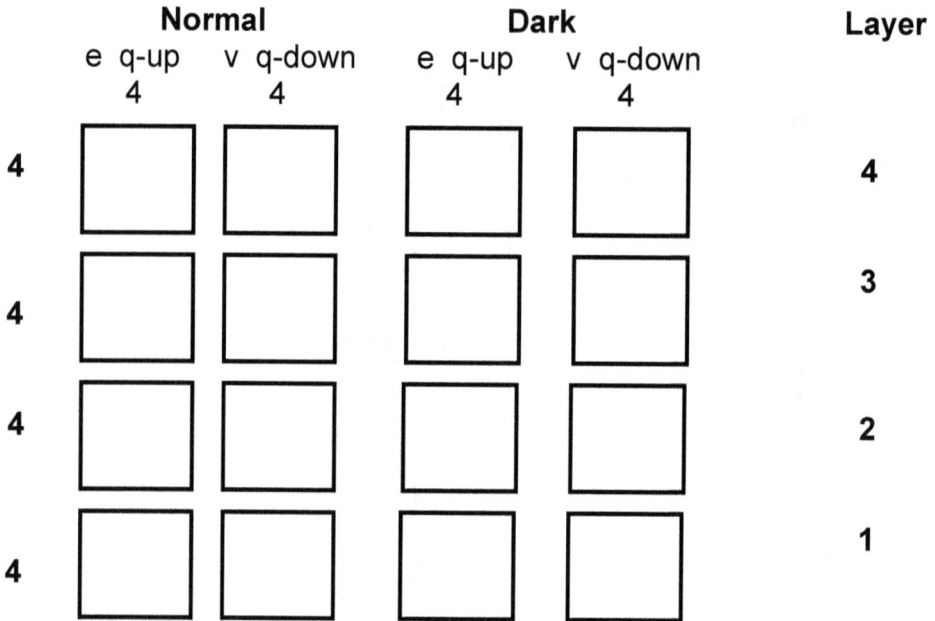

Figure 3.4. Block form of a 16 × 16 NEWQUeST/NEWUST fermion array with each block row corresponding to one layer. Each block contains four generations of fermions. The result is 4 × 4 blocks. The label e q-up indicates a charged lepton – up-type quark pair, v q-down indicates a neutral lepton – down-type quark pair, and so on. The blocks can be viewed as SU(3)⊗U(1) or broken SU(4) blocks. The three generations of fermions enclosed in a dotted rectangle are the 24 known fermions.

Appendix 3-A. Description of the NEWQUeST/NEWUST Internal Symmetry Groups

NEWUST and NEWQUeST contain sets of symmetry groups: Strong SU(3), Generation group U(4), Layer group U(4) and Connection group U(2). In this Appendix we describe their features. See Blaha (2019e) or (2020c) for more details on the Generation and Layer groups.

3-A.1 Strong SU(3)

The Strong Interaction Groups separately support rotations and interactions among the three up-type quarks and of the three down-type quarks in each generation, each layer, and in the Normal and Dark sectors. Each layer in the Normal and Dark sectors have a separate SU(3) group. Thus NEWQUeST/NEWUST each have a total of eight SU(3) Strong Interaction groups as shown in Figs. 3.1 and 3.2.

The fermions in each layer of the Normal and Dark sectors are different. Thus there are eight sets of fermions in the two block pairs in Fig. 3.4.

3-A. 2 U(4) Generation Groups

In the Big Bang all particles were massless and all symmetries unbroken. Hence the four Normal particle number symmetries, and the four Dark particle number symmetries, are all "conserved" in the Big Bang. Afterwards conservation are then broken afterwards in most cases.

We define two particle number operators for normal up-quark particles and down-quark particles, B_{uq} and B_{dq}. Similarly we define two particle number operators for normal species "e" (electron) particles and species "ν" particles, B_e and B_ν. Similarly we define Dark matter equivalents:[13] B_{De}, $B_{D\nu}$, B_{Duq}, and B_{Ddq}.

In the absence of symmetry breaking these fermion particle number operators would be conserved. Thus there are two sets of "diagonal" operators with associated U(4) groups for the Normal and Dark sectors. They are part of the Normal U(4) Generation Group and the Dark U(4) Generation Group.

The fermion fundamental representation of a U(4) group has four fermions. U(4) has rotations, and also interactions of the form $\overline{\Psi}\gamma \cdot B \cdot T\Psi$ where Ψ is a fermion four-vector, B is a 16 component U(4) gauge field, and T consists of 16 component 4×4 U(4) arrays.

In the case of the Generation Group the gauge fields have electric charge zero. Since the four species have different electric charges (1, 0, 2/3, -1/3) the U(4) gauge

---

[13] By analogy, we assume that there are four species of Dark matter: charged Dark leptons, neutral Dark leptons, Dark up-type quarks, and Dark down-type quarks. Thus we are led to the Dark particle numbers: Dark Baryon Numbers, and Dark Lepton Numbers shown above.

boson fields cannot mix the fermions of different species. Generation Group interactions are diagonal[14] in fermion species (e, ν, up-quark, and down-quark species).

Consequently the U(4) Generation Group must have a reducible representation D consisting of a set of four fundamental U(4) representations, D_e, D_ν, D_{upq}, and D_{dnq}, appearing in blocks along the diagonal of D. Each block is a separate U(4) irreducible representation for a species (due to the electric charge superselection rule.)

There are four generations of each species in the Normal and in the Dark matter sectors. The four generations for each fermion species: e, ν, up-quark, and down-quark each furnish a U(4) fundamental representation within the reducible representation D. The fourth generation of normal fermions has not as yet been found due to their extremely large masses.

The Generation Group rotates the fundamental fermions of each fundamental representation separately for each of the four species of each of the four layers.[15] Thus the Generation Group guarantees that all generations of each species have the same electric charge and other quantum numbers.

The U(4) Generation Group also specifies a gauge field interaction among the fermions of its fundamental representation, species by species, for both Normal and Dark sectors. The form of the interactions for the Normal sector for each fermion layer is:

$$g_e \overline{\Psi}_e \gamma \cdot B_e \cdot T\Psi_e + g_\nu \overline{\Psi}_\nu \gamma \cdot B_\nu \cdot T\Psi_\nu + g_{upq} \overline{\Psi}_{upq} \gamma \cdot B_{upq} \cdot T\Psi_{upq} + g_{dnq} \overline{\Psi}_{dnq} \gamma \cdot B_{dnq} \cdot T\Psi_{dnq}$$

$$(3\text{-}A.1)$$

where g_e ... are coupling constants, the gauge vector fields are B_e ... , and the Ψ_e ... are 4-vectors of fermions of the four generations of each species in a layer.

The gauge vector bosons of the Generation Group have large masses. If the conservation of the fermion particle numbers is broken then we view it as a consequence of Generation Group symmetry breaking.

Generation Group rotations guarantee the internal quantum numbers of each generation of each species are the same since symmetry breakdown is not present at the instant of the Big Bang.

The above discussion applies similarly to the Dark sector. There are 8 Generation Groups in total in NEWQUeST/NEWUST. See Figs. 3.1 and 3.2.

3-A.3 U(4) Layer Groups

The set[16] of particle number operators can be extended if we take account of the fourfold fermion generations.

We can subdivide the above particle number sets into four additional particle numbers *per generation*. For the ith generation (of the four generations) we define

---

[14] ElectroWeak interactions can cross between species due to their charged gauge vector bosons.

[15] There are separate Generation groups for each layer.

[16] Here again, in the Big Bang all particles were massless and all symmetries unbroken. Hence particle numbers are "conserved" in the Big Bang. Conservation is then broken afterwards in most cases.

L_{ie} – The "e" species particle number for the ith generation

L_{iv} – The v species particle number for the ith generation

L_{iuq} – The up-quark species particle number for the ith generation

L_{idq} – The down-quark species particle number for the ith generation

L_{iDe} – The Dark "e" species particle number for the ith generation

L_{iDv} – The Dark v species particle number for the ith generation

L_{iDuq} – The Dark up-quark species particle number for the ith generation

L_{iDdq} – Dark down-quark species particle number for the ith generation

for each generation i = 1, 2, 3, 4. Individual fermions have positive L_{ia} = +1 values and antifermions have negative L_{ia} = –1 values for each species.

At this point we have a set of four particle number operators for each of four generations (i = 1, 2, 3, 4) of fermions in the Normal sector and similarly in the Dark sector. We then define a U(4) group framework for each set of particle numbers.

The only way to specify fundamental representations for each of the four sets in a sector is to assume there are four layers, with each layer having four generations, and with a fundamental U(4) representation defined for each generation composed of fermions from each layer. Thus there are four Layer Groups for each Normal and each Dark sector: a Layer Group for generation 1, a Layer Group for generation 2, and so on.

The Layer Groups are also "split" by species due to the electric charge superselection rule. Each Layer Group is diagonal in the four fermion species. All their gauge fields are electrically neutral.

Consequently each of the four U(4) Layer Groups in the Normal fermion sector has a reducible U(4) representation D_j for j = 1, 2, 3, 4. Each reducible representation is composed of four irreducible U(4) representations for each species due to the electric charge superselection rule.:

$$D_j = D_{je} + D_{jv} + D_{jupq} + D_{jdnq},$$

for j = 1, 2, 3, 4.

There are four layers of each species in the Normal and in the Dark matter sectors. The second, third and fourth layers of normal fermions has not as yet been found due to their extremely large masses.

A Layer Group rotates the fundamental fermions of each fundamental representation separately for each of the four species of each of the four generations.

The Layer Groups guarantee that all layers of each species have the same electric charge and other quantum numbers.

Each U(4) Layer Group also specifies a gauge field interaction among the fermions of its fundamental representation, species by species, for both Normal and Dark sectors. The form of the interactions is:

$$g_{ei}\overline{\Psi}_{ei}\gamma \cdot C_{ei} \cdot T\Psi_{ei} + g_{vi}\overline{\Psi}_{vi}\gamma \cdot C_{vi} \cdot T\Psi_{vi} + g_{upqi}\overline{\Psi}_{upqi}\gamma \cdot C_{upqi} \cdot T\Psi_{upqi} + g_{dnqi}\overline{\Psi}_{dnqi}\gamma \cdot C_{dnqi} \cdot T\Psi_{dnqi}$$

$$(3\text{-}A.2)$$

for i = generation = 1, ... , 4, where g_{ei} ... are coupling constants, the gauge fields are C_{ei} ... , and the Ψ_{ei} ... are 4-vectors of fermions formed of the ith generation fermions in each layer of each species.

The gauge vector bosons of the Layer Groups also have large masses. If the conservation of the fermion particle numbers is broken then we view it as a consequence of Layer Groups symmetry breaking.

Layer Group rotations guarantee the internal quantum numbers of each layer of each species are the same since symmetry breakdown is not present at the instant of the Big Bang.

The above discussion applies similarly to the Dark sector. There are 8 of Layer Groups in NEWQUeST/NEWUST. See Figs. 3.1 and 3.2.

Fig. 3-A.1 shows the fundamental fermion spectrum with the representations of the Generation groups and Layer groups indicated.

Experimentally, we know of three generations of fermions—the lowest 3 generations of the lowest level. The remaining 4th generation and three layers of fermions are of much higher mass and are yet to be found.

See Blaha (2019g) and (2018e) for a detailed discussion of the Layer Groups. We note in passing that the symmetries of these number operators are badly broken. Yet the underlying group structure remains.

3-A.4 U(2) Connection Groups

The seven U(2) Connection groups of Fig. 3.3 generate rotations and interactions between corresponding fermions and vector bosons of each pair of blocks of the eight blocks of fermions in NEWQUeST/NEWUST.

Horizontal Lines

The horizontal lines in Fig. 3.3 (A, B, C, and D) each represent a U(2) Connection group that rotates two corresponding fermions in the Normal and Dark sectors of each layer. Thus a Normal e is rotated with a corresponding Dark e, and so on.

Each of the four horizontal Connection Groups has a reducible U(2) representation D that is the sum of 32 irreducible U(2) representations. We may view each reducible representation D as an array of 32 U(2) irreducible representations D_j strung along the diagonal.

$$D = \sum_{j=1}^{32} D_j$$

for each of the U(2) groups of the four horizontal lines in Fig. 3.3.

The U(2) group also specifies gauge field interactions between corresponding fermions in each layer of the Normal and Dark sectors of the form

$$g\overline{\Psi}_{Nn}\gamma{\cdot}A{\cdot}T\Psi_{Dn} \tag{3-A.3}$$

where N indicates a Normal fermion and D indicates the corresponding Dark fermion, with A being a U(2) gauge vector boson, and n the label for corresponding fermions.

These U(2) transformations imply that the Normal and Dark sectors have the same species and the same set of internal symmetries fermion by fermion.

Vertical Lines

The pairs of vertical lines in Fig. 3.3 (E, F, G) each represent a U(2) Connection group that rotates sets of two corresponding fermions in adjacent layers as shown in Fig. 3.3 in the Normal and Dark sectors. Thus a Normal e in layer 1 is rotated with a corresponding Normal e in layer 2, and so on.

Each of the three (six counting both Normal and Dark lines in Fig. 3.3) vertical Connection Groups has a reducible U(2) representation D that is the sum of 64 irreducible U(2) representations. We may view each reducible representation D as an array of 64 U(2) irreducible representations D_j strung along the diagonal.

$$D = \sum_{j=1}^{64} D_j$$

for each of the U(2) groups of the 3 (6) horizontal lines in Fig. 3.3. Note the 64 irreducible representations include both Normal and Dark sectors of a layer. [17]

Each U(2) group also specifies a gauge field interaction between corresponding fermions in adjacent layers for both Normal and Dark sectors:

$$g\overline{\Psi}_{nl_1}\gamma \cdot A \cdot T\Psi_{nl_2} \qquad\qquad (3\text{-}A.4)$$

where l_1 and l_2 designate layers, A is a gauge field vector boson, and n the label for corresponding fermions.

Each E, F, and G U(2) group reducible representation includes both Normal and Dark sectors.

3-A.5 A Unification of Symmetries in the NEWQUeST/NEWUST Fermion Spectrum

The Generation and Layer Groups are diagonal in the four fermion species. Fermions, considered species by species for all generations and layers in Normal and Dark sectors, have the same set of symmetries. Thus the Standard Model symmetries carry over directly to all parts of NEWQUeST/NEWUST.

If the Layer groups and the Connection groups were not present then each of the eight blocks may have differing sets of internal symmetries.

---

[17] There are 64 fermions in total for each of the four layers of NEWQUeST/NEWEST.

The Fermion Periodic Table

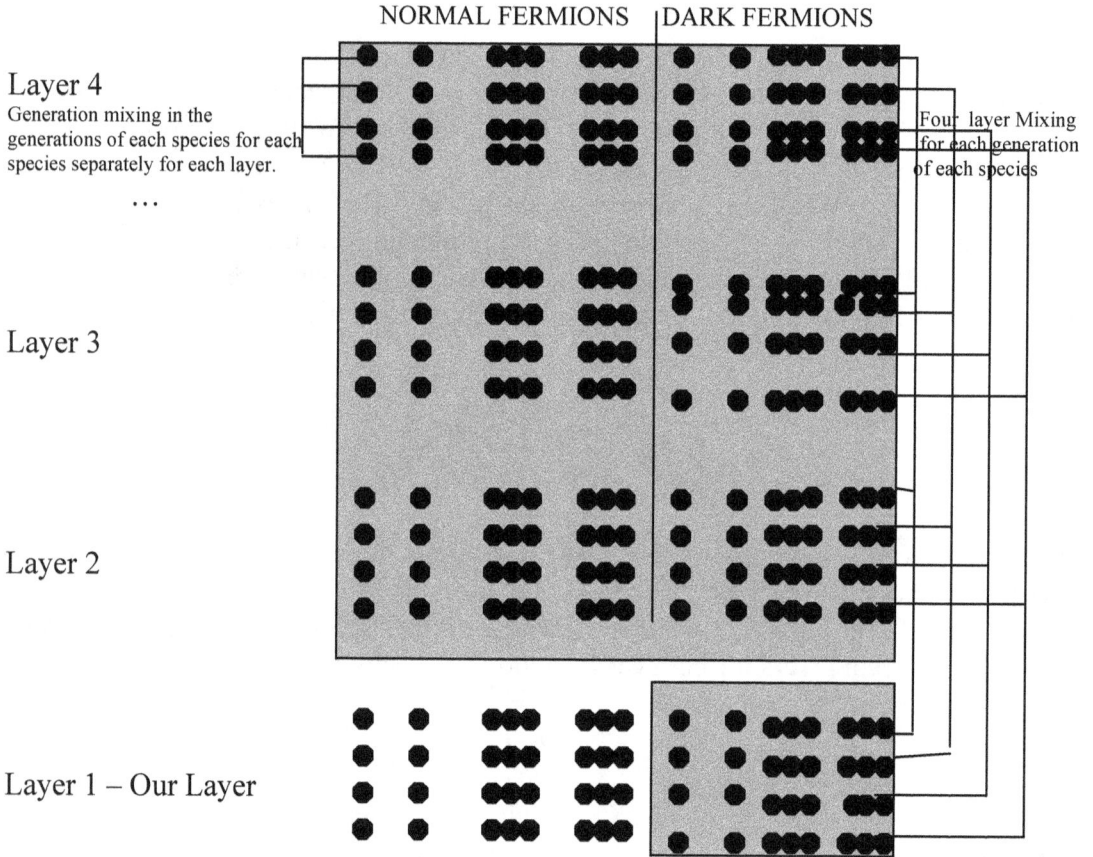

Figure 3-A.1. Fermion particle spectrum and partial examples of the pattern of mass mixing of the Generation groups and of the Layer groups. Unshaded parts are the known fermions with an additional, as yet not found, 4th generation. The lines on the left side (only shown for one layer) display the Generation mixing within each layer. The Generation mixing occurs within each layer using a separate Generation group for each layer. The lines on the right side show Layer group mixing (for Dark matter) with the mixing among all four layers for each of the four generations individually. There are four Layer groups for Normal matter and four Layer groups for Dark matter.. There are 256 fundamental fermions. NEWQUeST/NEWUST have the same fermion spectrum.

4. NEWUTMOST (for the Megaverse)

In defining the UST, the author was motivated to add a Megaverse within which the universe resided. Based on the form of UST, and the form of QUeST, the UTMOST octonion space was defined. Subsequently, after finding the origin of QUeST in Cayley numbers and developing the Octonion Cosmology Spectrum described in chapter 1 and earlier books in 2020, UTMOST was found to be the Cayley number n = 5 space. See Fig. 1.5.

Consequently, UTMOST can be viewed as consisting of four copies of QUeST. Thus due to the nature of Cayley numbers and the Octonion Spaces Spectrum, the total number of UTMOST/NEWUTMOST dimensions was 4 · 256 = 1024 dimensions as indicated in Fig. 1.5. The UTMOST/NEWUTMOST array was 32 × 32 in size and consists of four copies of the QUeST array. See Fig. 4.1. The internal symmetries of UTMOST/NEWUTMOST are four copies of Figs. 3.1 and 3.2.

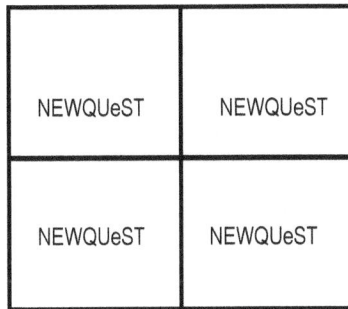

NEWQUeST	NEWQUeST
NEWQUeST	NEWQUeST

Figure 4.1. The UTMOST (and NEWUTMOST) dimension array viewed as composed of four copies of NEWQUeST.

The 32 space-time dimensions (four sets of four space-time dimensions from the four copies) can be combined to yield an UTMOST/NEWUTMOST U(4) Connection Group plus an eight real dimension space-time. Thus the total dimensions of the NEWUTMOST array is composed of four NEWQUeST sets of 252 internal symmetry dimensions[18] plus eight U(4) dimensions plus eight real space-time dimensions of NEWUTMOST giving a total of 1024 NEWUTMOST dimensions:

4 · 252 → 1008 + 8 U(4) dimensions = 1016   NEWUTMOST internal symmetry dimensions
                   + 8   NEWUTMOST real space-time dimensions
                 1024   Total NEWUTMOST dimensions

---

[18] NEWQUeST has four real space-time dimensions. These four NEWQUeST copies give 16 real space-time dimensions to NEWUTMOST. Eight of these 16 dimensions are transferred to form a U(4) Connection Group. The remainder becomes the 8 NEWUTMOST real space-time dimensions.

The NEWUTMOST (and UTMOST) dimension array may be decomposed into blocks of 16 dimensions as shown in Fig. 4.2. Pairs of two blocks (32 dimensions) form Dimension-32 Atoms as described in earlier books. We go beyond the Normal-Dark sectors of NEWUST defining various additional "Dark" sectors: Normal+Dark1, Dark2+Dark3, Dark4+Dark5 and Dark6+Dark7 as in Fig. 4.2.

Fig. 4.3 shows the breakdown of *one* NEWUTMOST layer into internal symmetry groups.

4.1.1 NEWUTMOST Internal Symmetries

The total internal symmetry is

$$\{[SU(2)\otimes U(1)\otimes SU(3)\otimes U(1)]^8\otimes U(4)^{16}\otimes U(2)^7\}^4\otimes U(4) \qquad (4.1)$$

Fig. 4.6 shows the four NEWQUeST parts of NEWUTMOST with the U(4) Connection Group. This group generates rotations among corresponding fermions (and gauge vector bosons) in the four parts. The group representation D is reducible. It has 256 irreducible fundamental U(4) groups D_i.

$$D = 256D_i \qquad (4.2)$$

4.1.2 Space-Time Coordinates

Eight real space-time coordinates.

4.2 Fundamental Fermion Spectrum

There are $4 \cdot 256 = 1024$ fundamental fermions in NEWUTMOST. See Figs. 4.4 and 4.5. Fig. 4.5 shows the fermion spectrum in a Strong interaction SU(4) format. It could be changed to an SU(3)⊗U(1) format.

4.3 Total Dimensions

The total of internal symmetry and space-time dimensions is 1024.

Normal + Dark1		Dark2 + Dark3		Dark4 + Dark5		Dark6 + Dark7	
4	4	4	4	4	4	4	4

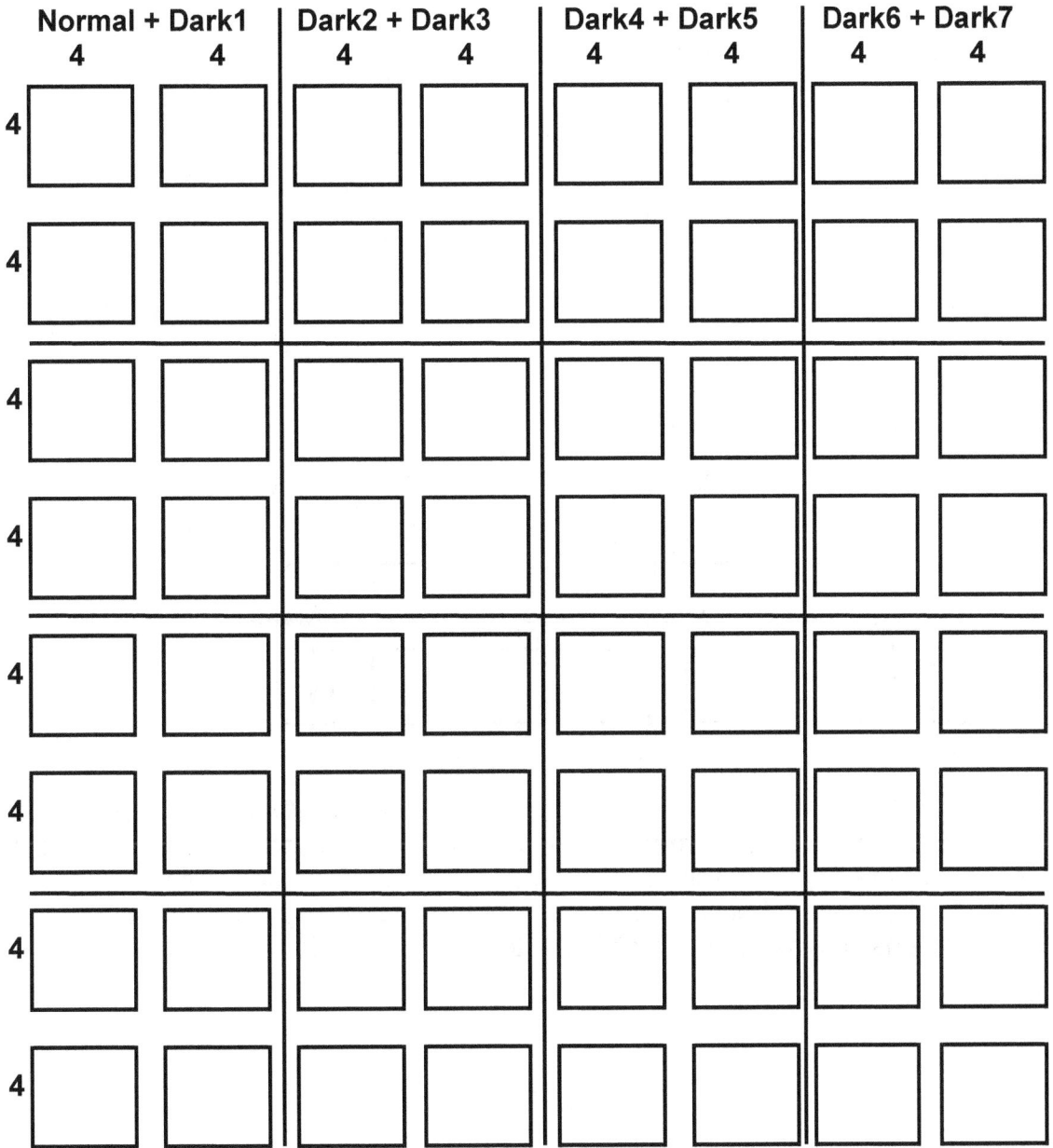

Figure 4.2. Four layers (each in two rows) in the 32 × 32 dimension NEWUTMOST array displayed as 4 × 4 = 16 dimension blocks counting space-time dimensions, some of which are later transferred to Connection groups. Each pair of these blocks form a Dimension-32 Atom. Two Dimension-32 Atoms form a 64 dimension blocks. Four Dimension-32 blocks form a 256 dimension layer.. The Dimension-32 Atoms pairs can be called the sections: Normal+Dark1, Dark2+Dark3, Dark4+Dark5 and Dark6+Dark7 as shown above. In total they form the 32 × 32 = 1024 NEWUTMOST dimension array.

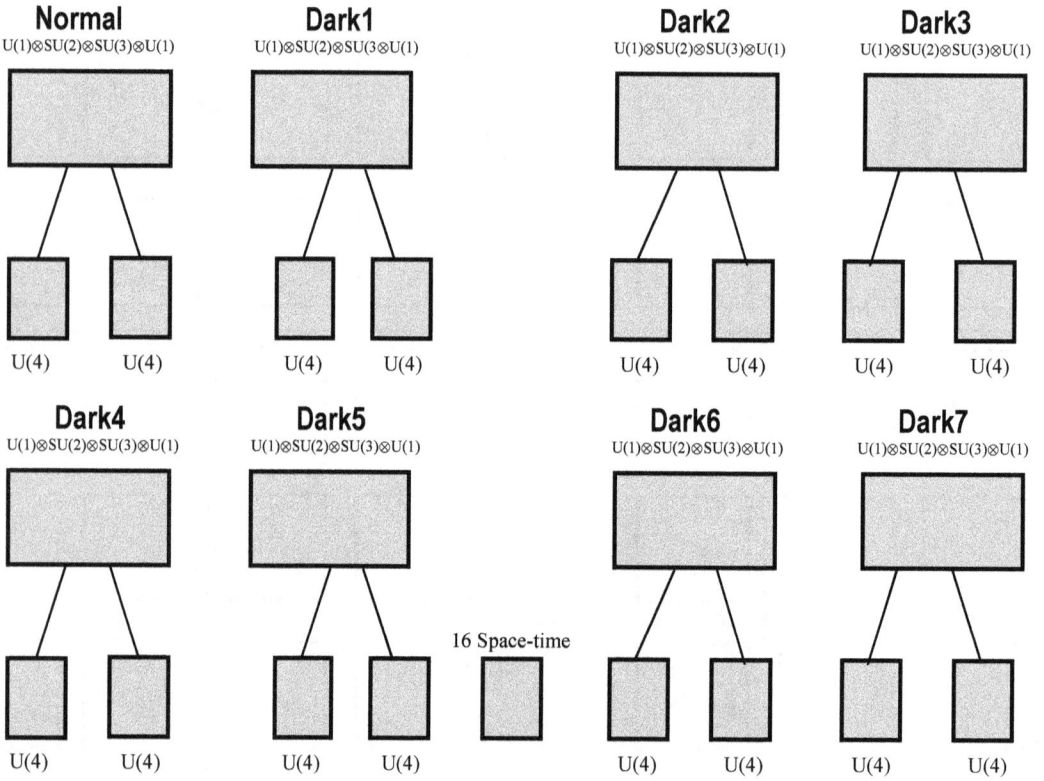

Figure 4.3. The internal symmetry groups of *one NEWUTMOST layer* of the four layers of 32 × 32 dimension NEWUTMOST. The other three layers are copies of this layer. Each U(1)⊗SU(2)⊗SU(3)⊗U(1) "block" displayed above has 12 dimensions because 4 of the 16 dimensions of a 16-block are space-time dimensions, which are transferred to form Connection Groups.

UTMOST Fermion Array

Normal	Dark1	Dark2	Dark3	Dark4	Dark5	Dark6	Dark7
●●●●●●●●	●●●●●●●●	●●●●●●●●	●●●●●●●●	●●●●●●●●	●●●●●●●●	●●●●●●●●	●●●●●●●●
●●●●●●●●	●●●●●●●●	●●●●●●●●	●●●●●●●●	●●●●●●●●	●●●●●●●●	●●●●●●●●	●●●●●●●●
●●●●●●●●	●●●●●●●●	●●●●●●●●	●●●●●●●●	●●●●●●●●	●●●●●●●●	●●●●●●●●	●●●●●●●●
●●●●●●●●	●●●●●●●●	●●●●●●●●	●●●●●●●●	●●●●●●●●	●●●●●●●●	●●●●●●●●	●●●●●●●●
●●●●●●●●	●●●●●●●●	●●●●●●●●	●●●●●●●●	●●●●●●●●	●●●●●●●●	●●●●●●●●	●●●●●●●●
●●●●●●●●	●●●●●●●●	●●●●●●●●	●●●●●●●●	●●●●●●●●	●●●●●●●●	●●●●●●●●	●●●●●●●●
●●●●●●●●	●●●●●●●●	●●●●●●●●	●●●●●●●●	●●●●●●●●	●●●●●●●●	●●●●●●●●	●●●●●●●●
●●●●●●●●	●●●●●●●●	●●●●●●●●	●●●●●●●●	●●●●●●●●	●●●●●●●●	●●●●●●●●	●●●●●●●●
●●●●●●●●	●●●●●●●●	●●●●●●●●	●●●●●●●●	●●●●●●●●	●●●●●●●●	●●●●●●●●	●●●●●●●●
●●●●●●●●	●●●●●●●●	●●●●●●●●	●●●●●●●●	●●●●●●●●	●●●●●●●●	●●●●●●●●	●●●●●●●●
●●●●●●●●	●●●●●●●●	●●●●●●●●	●●●●●●●●	●●●●●●●●	●●●●●●●●	●●●●●●●●	●●●●●●●●
●●●●●●●●	●●●●●●●●	●●●●●●●●	●●●●●●●●	●●●●●●●●	●●●●●●●●	●●●●●●●●	●●●●●●●●
●●●●●●●●	●●●●●●●●	●●●●●●●●	●●●●●●●●	●●●●●●●●	●●●●●●●●	●●●●●●●●	●●●●●●●●
●●●●●●●●	●●●●●●●●	●●●●●●●●	●●●●●●●●	●●●●●●●●	●●●●●●●●	●●●●●●●●	●●●●●●●●
●●●●●●●●	●●●●●●●●	●●●●●●●●	●●●●●●●●	●●●●●●●●	●●●●●●●●	●●●●●●●●	●●●●●●●●
●●●●●●●●	●●●●●●●●	●●●●●●●●	●●●●●●●●	●●●●●●●●	●●●●●●●●	●●●●●●●●	●●●●●●●●

Figure 4.4. Spectrum of UTMOST fermions in a 16×64 format. Each fermion is represented by a ●..Each set of eight ●.'s represents a charged lepton, a neutral lepton, three up-type quarks, and three down-type quarks. There are eight sets of four species in four generations which are in turn in 4 layers. There are 1024 fundamental fermions taking account of quark triplets. The 16×64 format could be changed to a 32×32 format without adverse physical consequences by "juggling" the Dark sectors. See Fig. 4.5.

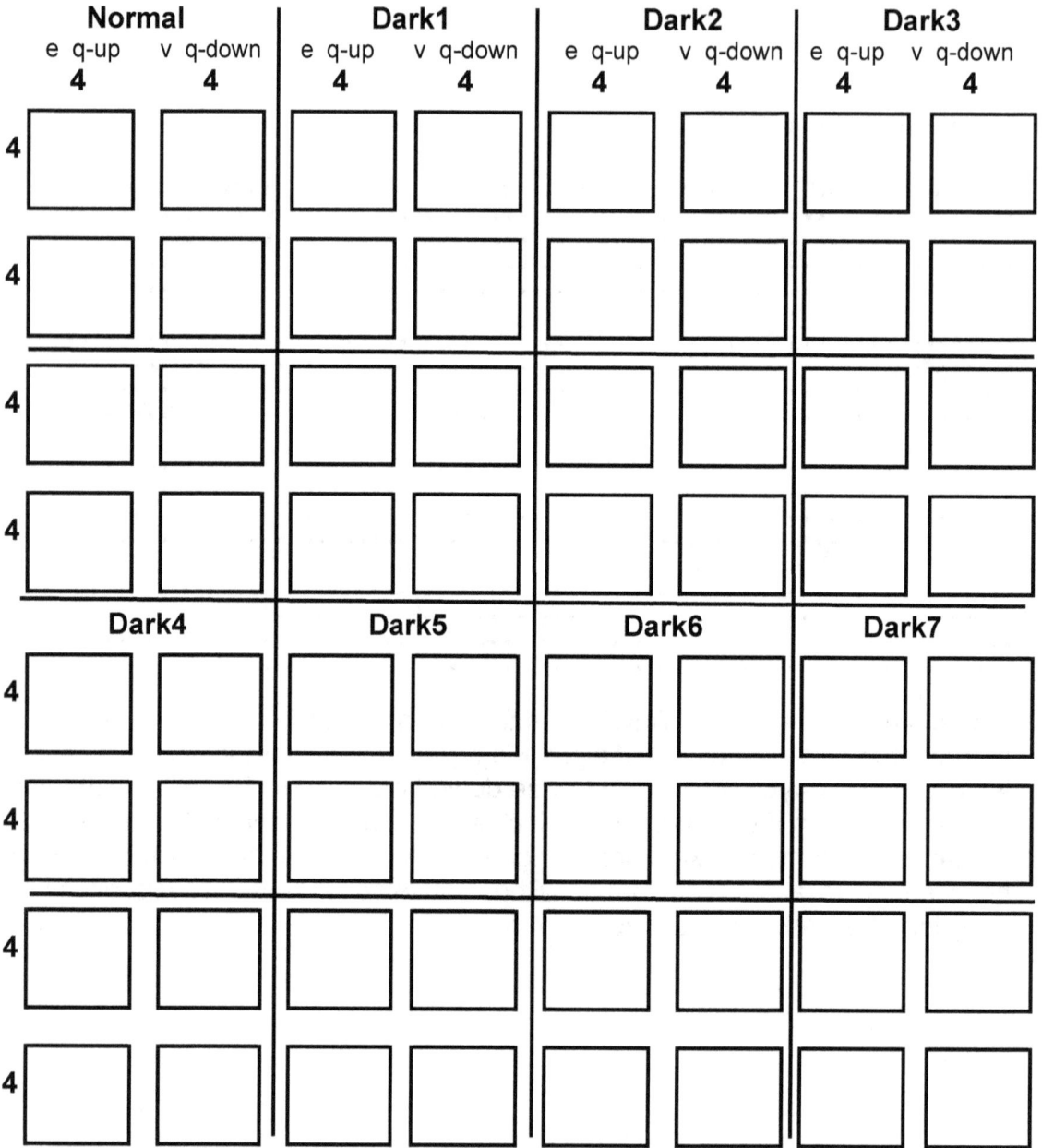

Figure 4.5. Block form of the 32 × 32 UTMOST fermion array with each row corresponding to *half of an UTMOST layer*. Thus 8 × ½ = 4 layers results. Each block contains four generations of fermions. The result is sixty-four 4 × 4 blocks. The label e q-up indicates a charged lepton – up-type quark pair, v q-down indicates a neutral lepton – down-type quark pair, and so on. *The form displayed here explains why generations come in fours.*

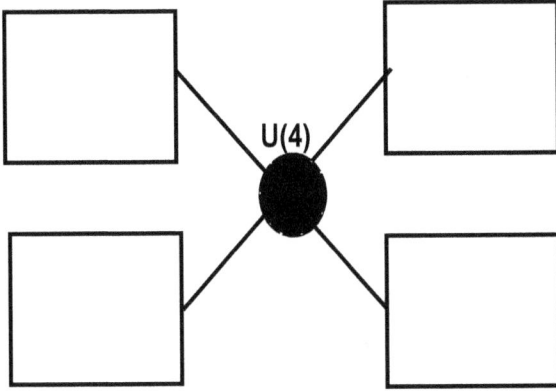

Figure 4.6. The U(4) Connection Group of NEWUTMOST.

5. The Octonion – DNA Parallel: Cosmic Indifference or Cosmic Propinquity?

"Nature repeats itself" is a familiar truism. Octonion Cosmology has some features that are very similar in structure to those of DNA-based life. The similarity, at the very least, gives one pause for thought.

One could view the similarity as accidental; or one could view the similarity as the result of design. We chose to suspend judgement on this matter and simply describe the analogous structures.

The resemblance of Octonion Cosmology features and life, in particular DNA-based life, cannot be viewed as a justification of Octonion Cosmology (OC) although the appearance of life in the universe seems to follow naturally from OC.

In this chapter we will describe the similarity between NEWQUeST OC features and Life features as evidenced by the form of DNA.[19,20]

5.1 The Structure of OC NEWQUeST

Chapter 3 contains a schematic for NEWQUeST. The form of NEWQUeST for both symmetry groups and for fermions is displayed in Fig. 5.1. It has four layers for both Normal and Dark sectors of groups and fermions. The layers in both the Normal and Dark sectors are interconnected by seven U(2) groups. Each group specifies an interaction between the corresponding fermions of blocks. For example an electron in the top Normal layer has an interaction with a Dark electron in the top Dark layer, and so on.

The seven interactions must be ultraweak at current accelerator energies and their corresponding vector bosons must be extremely massive.

5.1.1 Rules Governing the NEWQUeST Structure

The form of the NEWQUeST dimension array is the result of certain rules within the definition of Octonion spaces:

1. The array must be a square array of 16 by 16 dimensions implying a 16 by 16 array of fundamental symmetry group dimensions (composed of 32-Dimension Atoms) and a 16 by 16 array of fundamental fermions (composed of 32-Fermion Atoms)

2. There is a Normal and a Dark sector.

---

[19] The author was one of the early investigators of DNA's electronic properties. In 1964 he performed Electron Spin Resonance (ESR) studies of DNA, and its bases: Adenine, Thymine, Guanine, and Cytosine at Wayne State University (unpublished).
[20] DNA-related structures such as Guanine-Tetrads (G-tetrads) and G-quadruplexes also have analogues in OC spaces.

3. Each sector has four layers of equal size. Each Normal and Dark layer is a 32-Dimension Atom for symmetries. Each Normal and Dark layer of fermions is a 32-Fermion Atom. A 32-Fermion Atom has four generations of fermions with each fermion generation having an e, ν, 3 up-type quarks, and 3 down-type quarks.

4. There are four U(2) groups providing interactions between corresponding Normal and Dark blocks for each layer. (Fig. 5.1)

5. There are three U(2) groups providing interactions vertically between fermions in adjacent Normal layers. These same U(2) groups provide interactions vertically between fermions in adjacent Dark layers. (Fig. 5.1)

These rules fix the form of the NEWQUeST arrays for internal symmetries and fermions.

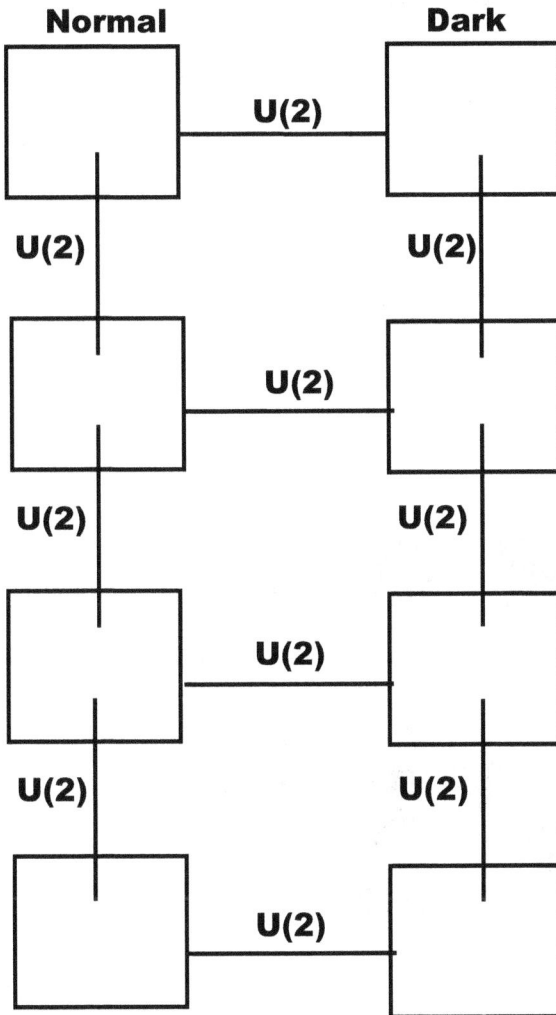

Figure 5.1. Schmatic for NEWQUeST.

5.2 The Structure of DNA

DNA is the basis of all known multicellular life on earth. Based on the principle that the earth is not "special", and the tendency of Nature to repeat itself, it is reasonable to view all life in the universe as based on DNA or very similar analogues. Carbon is abundant in the universe. Due to nuclear fusion in the stars it is about ten times more common than silicon, which is often touted as a carbon alternative. Carbon has chemical features that make it especially suitable for composing life.

DNA consists of a helix composed of two intertwined strands of four base pair fragments. A four base pair fragment of DNA is depicted in Fig. 5.2.

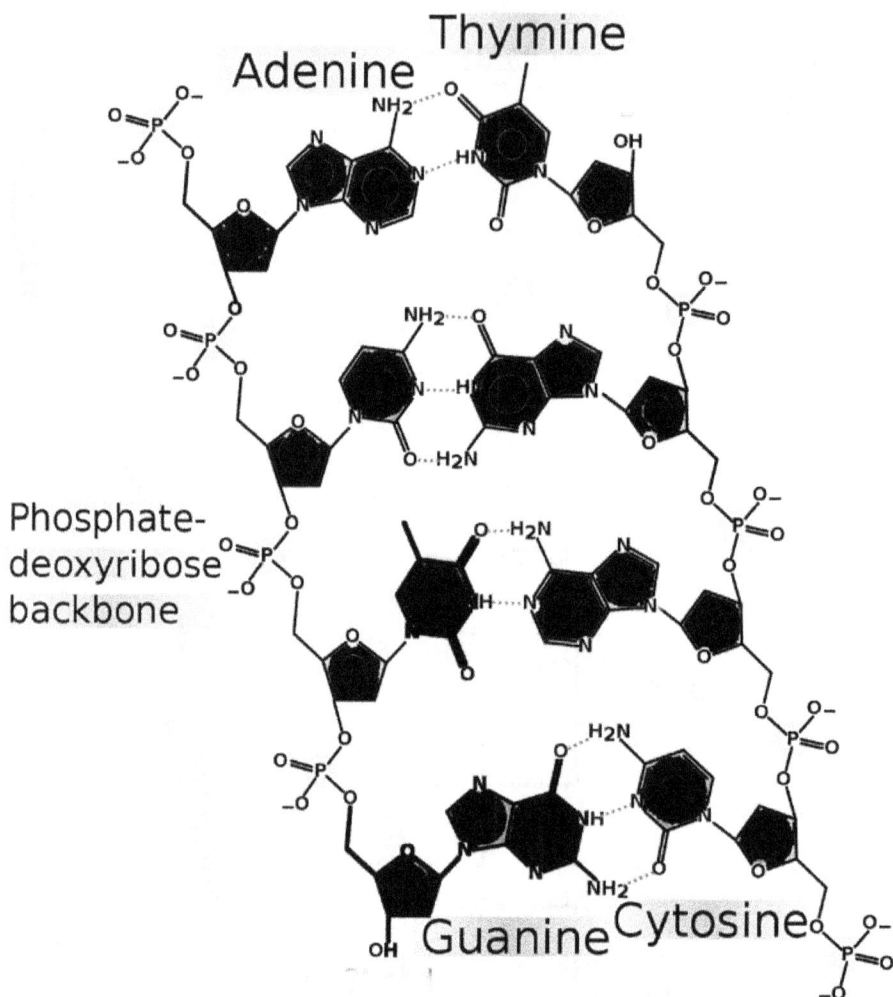

Figure 5.2. A four base pair fragment of double helix DNA. The bases are Adenine, Thymine, Guanine, and Cytosine. DNA is composed of many such fragments.

5.2.1 ""Possible" DNA Analogues

Alternatives to DNA have been studied. They replace elements in DNA-like strands with phosphorus, silica, arsenic, metals, and so on. None of these possibilities have the flexibility of DNA for life processes. Some of the DNA alternatives that have been studied are:

Hachimoji DNA
PNA, which supports triple helix strands
And LNA, GNA, HNA, TNA

5.3 NEWQUeST Analogue of DNA

In this section we superimpose the structure of NEWQUeST on a DNA fragment. The comparison clearly shows the analogous structure of DNA fragments and NEWQUeST dimension space.

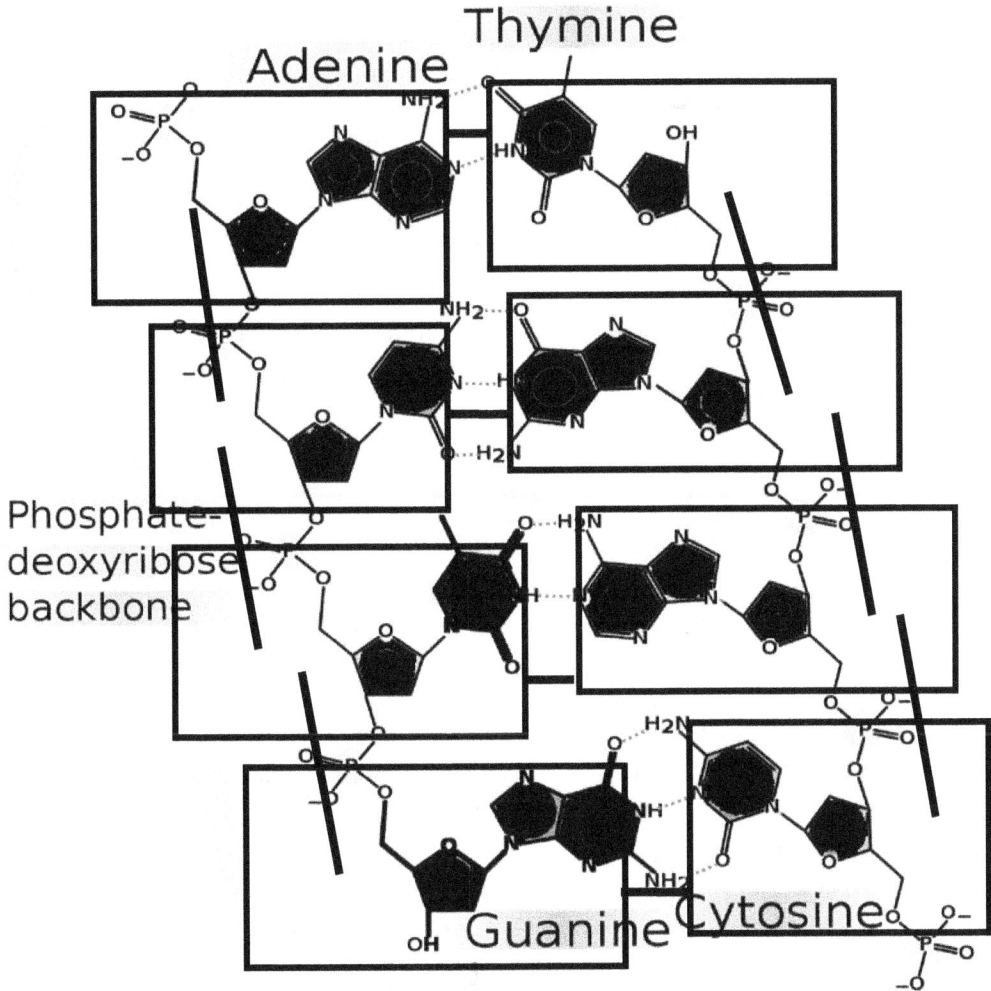

Figure 5.3. The NEWQUeST structure superimposed on a four base pair fragment of double helix DNA. The correspondence is evident. Four hydrogen bonds between strands corresponds to four U(2) interactions (very short horizontal lines) between "Normal" and Dark sectors. The three vertical lines link NEWQUeST layers with U(2) interactions. They correspond to phospho-diester covalent bonds linking nucleotides to each other. Nucleotides are composed of the bases: Adenine, Thymine, Guanine, and Cytosine.

5.4 Heuristic Relation of Life from OC

The analogues discussed above raise the question: Are they the result of Chance or do they reflect a deeper Reality? In this section we will suggest that they are the result of a deeper Reality by outlining a heuristic relation of Life (DNA) to Octonion Cosmology. The deeper reality is centered on The Standard Model sector of Octonion Cosmology.

The ingredients of Life revolve around the structure of DNA – the primary ingredient of all multicellular life... The major factors determining its structure are the electromagnetism of the electrons within DNA and the properties of Carbon.[21] Thus the key factors that lead to Life are Electrodynamics and the Nuclear Force generated by the Strong Interaction. The Nuclear Force favors Carbon production in the stars. Comparing Carbon abundance to Silicon abundance, in particular, we see Carbon is favored by a 10:1 factor.

Two interactions (forces) provide the critical basis for DNA and Life as we know it: Quantum Electrodynamics and the Strong Interaction.

Turning now to our NEWQUeST and Unified SuperStandard Theory (UST) we find the following facts:

1. NEWQUeST and UST are built from Dimension-32 Atoms that are relatively isolated because the forces between them are weak due to large mass gauge fields and symmetry breaking.

2. The one Dimension-32 Atom, with which we are familiar, is that of The Standard Model. It contains the Strong Interaction and Electromagnetism.

3. Thus there is a direct connection between Octonion Cosmology and Life through the Standard Model sector of Octonion Cosmology. Remarkably the similar structure of DNA fragments and NEWQUeST is coincidental but intriguing.

4. The value of the Fine Structure Constant, whose specific value is critical for Life, has been shown to be the same at all times and locations in all possible universes. See Blaha (2019f). The value is a consequence of vacuum polarization in QED.

The connection between OC and Life has cautionary points:

A. The detailed calculation of the electronic structure of DNA would be difficult.

---

[21] Other elements play a role also But the molecular flexibility of Carbon appears critical.

B. The existence of DNA-based Life does not prove OC since the relation boils down to The Standard Model. OC leads to The Standard Model as do other proposed fundamental theories.

5.5 The OC – Life Paradigm

We see a parallel between OC and DNA:

NEWQUeST $\rightarrow$ DNA
U(2) $\rightarrow$ Hydrogen Bonds
U(2) $\rightarrow$ phospho-diester covalent bonds linking nucleotides

Thus a schematic-based paradigm may be said to exist.

Octonion Cosmology does not uniquely imply DNA-based Life. However, Octonion Cosmology is consistent with the existence of DNA-based Life. The match of the structure of DNA fragments with the structure of NEWQUeST is most suggestive.

REFERENCES

Akhiezer, N. I., Frink, A. H. (tr), 1962, *The Calculus of Variations* (Blaisdell Publishing, New York, 1962).

Bjorken, J. D., Drell, S. D., 1964, *Relativistic Quantum Mechanics* (McGraw-Hill, New York, 1965).

Bjorken, J. D., Drell, S. D., 1965, *Relativistic Quantum Fields* (McGraw-Hill, New York, 1965).

Blaha, S., 1995, *C++ for Professional Programming* (International Thomson Publishing, Boston, 1995).

_____, 1998, *Cosmos and Consciousness* (Pingree-Hill Publishing, Auburn, NH, 1998 and 2002).

_____, 2002, *A Finite Unified Quantum Field Theory of the Elementary Particle Standard Model and Quantum Gravity Based on New Quantum Dimensions™ & a New Paradigm in the Calculus of Variations* (Pingree-Hill Publishing, Auburn, NH, 2002).

_____, 2004, *Quantum Big Bang Cosmology: Complex Space-time General Relativity, Quantum Coordinates™ Dodecahedral Universe, Inflation, and New Spin 0, ½, 1 & 2 Tachyons & Imagyons* (Pingree-Hill Publishing, Auburn, NH, 2004).

_____, 2005a, *Quantum Theory of the Third Kind: A New Type of Divergence-free Quantum Field Theory Supporting a Unified Standard Model of Elementary Particles and Quantum Gravity based on a New Method in the Calculus of Variations* (Pingree-Hill Publishing, Auburn, NH, 2005).

_____, 2005b, *The Metatheory of Physics Theories, and the Theory of Everything as a Quantum Computer Language* (Pingree-Hill Publishing, Auburn, NH, 2005).

_____, 2005c, *The Equivalence of Elementary Particle Theories and Computer Languages: Quantum Computers, Turing Machines, Standard Model, Superstring Theory, and a Proof that Gödel's Theorem Implies Nature Must Be Quantum* (Pingree-Hill Publishing, Auburn, NH, 2005).

_____, 2006a, *The Foundation of the Forces of Nature* (Pingree-Hill Publishing, Auburn, NH, 2006).

_____, 2006b, *A Derivation of ElectroWeak Theory based on an Extension of Special Relativity; Black Hole Tachyons; & Tachyons of Any Spin.* (Pingree-Hill Publishing, Auburn, NH, 2006).

_____, 2007a, *Physics Beyond the Light Barrier: The Source of Parity Violation, Tachyons, and A Derivation of Standard Model Features* (Pingree-Hill Publishing, Auburn, NH, 2007).

_____, 2007b, *The Origin of the Standard Model: The Genesis of Four Quark and Lepton Species, Parity Violation, the ElectroWeak Sector, Color SU(3), Three Visible Generations of Fermions, and One Generation of Dark Matter with Dark Energy* (Pingree-Hill Publishing, Auburn, NH, 2007).

_____, 2008a, *A Direct Derivation of the Form of the Standard Model From GL(16)* (Pingree-Hill Publishing, Auburn, NH, 2008).

50   **REFERENCES**

_____, 2008b, *A Complete Derivation of the Form of the Standard Model With a New Method to Generate Particle Masses Second Edition* (Pingree-Hill Publishing, Auburn, NH, 2008)

_____, 2009, *The Algebra of Thought & Reality: The Mathematical Basis for Plato's Theory of Ideas, and Reality Extended to Include A Priori Observers and Space-Time Second Edition* (Pingree-Hill Publishing, Auburn, NH, 2009).

_____, 2010a, *Operator Metaphysics: A New Metaphysics Based on a New Operator Logic and a New Quantum Operator Logic that Lead to a Mathematical Basis for Plato's Theory of Ideas and Reality* (Pingree-Hill Publishing, Auburn, NH, 2010).

_____, 2010b, *The Standard Model's Form Derived from Operator Logic, Superluminal Transformations and GL(16)* (Pingree-Hill Publishing, Auburn, NH, 2010).

_____, 2010c, *SuperCivilizations: Civilizations as Superorganisms* (McMann-Fisher Publishing, Auburn, NH, 2010).

_____, 2011a, *21st Century Natural Philosophy Of Ultimate Physical Reality* (McMann-Fisher Publishing, Auburn, NH, 2011).

_____, 2011b, *All the Universe! Faster Than Light Tachyon Quark Starships & Particle Accelerators with the LHC as a Prototype Starship Drive Scientific Edition* (Pingree-Hill Publishing, Auburn, NH, 2011).

_____, 2011c, *From Asynchronous Logic to The Standard Model to Superflight to the Stars* (Blaha Research, Auburn, NH, 2011).

_____, 2012a, *From Asynchronous Logic to The Standard Model to Superflight to the Stars volume 2: Superluminal CP and CPT, U(4) Complex General Relativity and The Standard Model, Complex Vierbein General Relativity, Kinetic Theory, Thermodynamics* (Blaha Research, Auburn, NH, 2012).

_____, 2012b, *Standard Model Symmetries, And Four And Sixteen Dimension Complex Relativity; The Origin Of Higgs Mass Terms* (Blaha Reasearch, Auburn, NH, 2012).

_____, 2013a, *Multi-Stage Space Guns, Micro-Pulse Nuclear Rockets, and Faster-Than-Light Quark-Gluon Ion Drive Starships* (Blaha Research, Auburn, NH, 2013).

_____, 2013b, *The Bridge to Dark Matter; A New Sister Universe; Dark Energy; Inflatons; Quantum Big Bang; Superluminal Physics; An Extended Standard Model Based on Geometry* (Blaha Reasearch, Auburn, NH, 2013).

_____, 2014a, *Universes and Megaverses: From a New Standard Model to a Physical Megaverse; The Big Bang; Our Sister Universe's Wormhole; Origin of the Cosmological Constant, Spatial Asymmetry of the Universe, and its Web of Galaxies; A Baryonic Field between Universes and Particles; Megaverse Extended Wheeler-DeWitt Equation* (Blaha Reasearch, Auburn, NH, 2014).

_____, 2014b, *All the Megaverse! Starships Exploring the Endless Universes of the Cosmos Using the Baryonic Force* (Blaha Research, Auburn, NH, 2014).

_____, 2014c, *All the Megaverse! II Between Megaverse Universes: Quantum Entanglement Explained by the Megaverse Coherent Baryonic Radiation Devices – PHASERs Neutron Star Megaverse Slingshot*

Dynamics Spiritual and UFO Events, and the Megaverse Microscopic Entry into the Megaverse (Blaha Research, Auburn, NH, 2014).

_____, 2015a, *PHYSICS IS LOGIC PAINTED ON THE VOID: Origin of Bare Masses and The Standard Model in Logic, U(4) Origin of the Generations, Normal and Dark Baryonic Forces, Dark Matter, Dark Energy, The Big Bang, Complex General Relativity, A Megaverse of Universe Particles* (Blaha Research, Auburn, NH, 2015).

_____, 2015b, *PHYSICS IS LOGIC Part II: The Theory of Everything, The Megaverse Theory of Everything, U(4)⊗U(4) Grand Unified Theory (GUT), Inertial Mass = Gravitational Mass, Unified Extended Standard Model and a New Complex General Relativity with Higgs Particles, Generation Group Higgs Particles* (Blaha Research, Auburn, NH, 2015).

_____, 2015c, *The Origin of Higgs ("God") Particles and the Higgs Mechanism: Physics is Logic III, Beyond Higgs – A Revamped Theory With a Local Arrow of Time, The Theory of Everything Enhanced, Why Inertial Frames are Special, Universes of the Mind* (Blaha Research, Auburn, NH, 2015).

_____, 2015d, *The Origin of the Eight Coupling Constants of The Theory of Everything: U(8) Grand Unified Theory of Everything (GUTE), S^8 Coupling Constant Symmetry, Space-Time Dependent Coupling Constants, Big Bang Vacuum Coupling Constants, Physics is Logic IV* (Blaha Research, Auburn, NH, 2015).

_____, 2016a, *New Types of Dark Matter, Big Bang Equipartition, and A New U(4) Symmetry in the Theory of Everything: Equipartition Principle for Fermions, Matter is 83.33% Dark, Penetrating the Veil of the Big Bang, Explicit QFT Quark Confinement and Charmonium, Physics is Logic V* (Blaha Research, Auburn, NH, 2016).

_____, 2016b, *The Periodic Table of the 192 Quarks and Leptons in The Theory of Everything: The U(4) Layer Group, Physics is Logic VI* (Blaha Research, Auburn, NH, 2016).

_____, 2016c, *New Boson Quantum Field Theory, Dark Matter Dynamics, Dark Matter Fermion Layer Mixing, Genesis of Higgs Particles, New Layer Higgs Masses, Higgs Coupling Constants, Non-Abelian Higgs Gauge Fields, Physics is Logic VII* (Blaha Research, Auburn, NH, 2016).

_____, 2016d, *Unification of the Strong Interactions and Gravitation: Quark Confinement Linked to Modified Short-Distance Gravity; Physics is Logic VIII* (Blaha Research, Auburn, NH, 2016).

_____, 2016e, *MoND: Unification of the Strong Interactions and Gravitation II, Quark Confinement Linked to Large-Scale Gravity, Physics is Logic IX* (Blaha Research, Auburn, NH, 2016).

_____, 2016f, *CQ Mechanics: A Unification of Quantum & Classical Mechanics, Quantum/Semi-Classical Entanglement, Quantum/Classical Path Integrals, Quantum/Classical Chaos* (Blaha Research, Auburn, NH, 2016).

_____, 2016g, *GEMS: Unified Gravity, ElectroMagnetic and Strong Interactions: Manifest Quark Confinement, A Solution for the Proton Spin Puzzle, Modified Gravity on the Galactic Scale* (Pingree Hill Publishing, Auburn, NH, 2016).

_____, 2016h, *Unification of the Seven Boson Interactions based on the Riemann-Christoffel Curvature Tensor* (Pingree Hill Publishing, Auburn, NH, 2016).

_____, 2017a, *Unification of the Eleven Boson Interactions based on 'Rotations of Interactions'* (Pingree Hill Publishing, Auburn, NH, 2017).

_____, 2017b, *The Origin of Fermions and Bosons, and Their Unification* (Pingree Hill Publishing, Auburn, NH, 2017).

_____, 2017c, *Megaverse: The Universe of Universes* (Pingree Hill Publishing, Auburn, NH, 2017).

_____, 2017d, *SuperSymmetry and the Unified SuperStandard Model* (Pingree Hill Publishing, Auburn, NH, 2017).

_____, 2017e, *From Qubits to the Unified SuperStandard Model with Embedded SuperStrings: A Derivation* (Pingree Hill Publishing, Auburn, NH, 2017).

_____, 2017f, *The Unified SuperStandard Model in Our Universe and the Megaverse: Quarks, ... ,* (Pingree Hill Publishing, Auburn, NH, 2017).

_____, 2018a, *The Unified SuperStandard Model and the Megaverse SECOND EDITION A Deeper Theory based on a New Particle Functional Space that Explicates Quantum Entanglement Spookiness (Volume 1)* (Pingree Hill Publishing, Auburn, NH, 2018).

_____, 2018b, *Cosmos Creation: The Unified SuperStandard Model, Volume 2, SECOND EDITION* (Pingree Hill Publishing, Auburn, NH, 2018).

_____, 2018c, *God Theory (*Pingree Hill Publishing, Auburn, NH, 2018).

_____, 2018d, *Immortal Eye: God Theory: Second Edition* (Pingree Hill Publishing, Auburn, NH, 2018).

_____, 2018e, *Unification of God Theory and Unified SuperStandard Model THIRD EDITION* (Pingree Hill Publishing, Auburn, NH, 2018).

_____, 2019a, *Calculation of: QED α = 1/137, and Other Coupling Constants of the Unified SuperStandard Theory* (Pingree Hill Publishing, Auburn, NH, 2019).

_____, 2019b, *Coupling Constants of the Unified SuperStandard Theory SECOND EDITION* (Pingree Hill Publishing, Auburn, NH, 2019).

_____, 2019c, *New Hybrid Quantum Big_Bang–Megaverse_Driven Universe with a Finite Big Bang and an Increasing Hubble Constant* (Pingree Hill Publishing, Auburn, NH, 2019).

_____, 2019d, *The Universe, The Electron and The Vacuum* (Pingree Hill Publishing, Auburn, NH, 2019).

_____, 2019e, *Quantum Big Bang – Quantum Vacuum Universes (Particles)* (Pingree Hill Publishing, Auburn, NH, 2019).

REFERENCES **53**

_____, 2019f, *The Exact QED Calculation of the Fine Structure Constant Implies ALL 4D Universes have the Same Physics/Life Prospects* (Pingree Hill Publishing, Auburn, NH, 2019).

_____, 2019g, *Unified SuperStandard Theory and the SuperUniverse Model: The Foundation of Science* (Pingree Hill Publishing, Auburn, NH, 2019).

_____, 2020a, *Quaternion Unified SuperStandard Theory (The QUeST) and Megaverse Octonion SuperStandard Theory (MOST)* (Pingree Hill Publishing, Auburn, NH, 2020).

_____, 2020b, *United Universes Quaternion Universe - Octonion Megaverse* (Pingree Hill Publishing, Auburn, NH, 2020).

_____, 2020c, *Unified SuperStandard Theories for Quaternion Universes & The Octonion Megaverse* (Pingree Hill Publishing, Auburn, NH, 2020).

_____, 2020d, *The Essence of Eternity: Quaternion & Octonion SuperStandard Theories* (Pingree Hill Publishing, Auburn, NH, 2020).

_____, 2020e, *The Essence of Eternity II* (Pingree Hill Publishing, Auburn, NH, 2020).

_____, 2020f, *A Very Conscious Universe* (Pingree Hill Publishing, Auburn, NH, 2020).

_____, 2020g, *Hypercomplex Universe* (Pingree Hill Publishing, Auburn, NH, 2020).

_____, 2020h, *Beneath the Quaternion Universe* (Pingree Hill Publishing, Auburn, NH, 2020).

_____, 2020i, *Why is the Universe Real? From Quaternion & Octonion to Real Coordinates* (Pingree Hill Publishing, Auburn, NH, 2020).

_____, 2020j, *The Origin of Universes: of Quaternion Unified SuperStandard Theory (QUeST); and of the Octonion Megaverse (UTMOST)* (Pingree Hill Publishing, Auburn, NH, 2020).

_____, 2020k, *The Seven Spaces of Creation: Octonion Cosmology* (Pingree Hill Publishing, Auburn, NH, 2020).

_____, 2020l, *From Octonion Cosmology to the Unified SuperStandard Theory of Particles* (Pingree Hill Publishing, Auburn, NH, 2020).

_____, 2021a, *Pioneering the Cosmos* (Pingree Hill Publishing, Auburn, NH, 2021).

_____, 2021b, *Pioneering the Cosmos II* (Pingree Hill Publishing, Auburn, NH, 2021).

_____, 2021c, *Beyond Octonion Cosmology* (Pingree Hill Publishing, Auburn, NH, 2021).

_____, 2021d, *Universes are Particles* (Pingree Hill Publishing, Auburn, NH, 2021).

Eddington, A. S., 1952, *The Mathematical Theory of Relativity* (Cambridge University Press, Cambridge, U.K., 1952).

Fant, Karl M., 2005, *Logically Determined Design: Clockless System Design With NULL Convention Logic* (John Wiley and Sons, Hoboken, NJ, 2005).

Feinberg, G. and Shapiro, R., 1980, *Life Beyond Earth: The Intelligent Earthlings Guide to Life in the Universe* (William Morrow and Company, New York, 1980).

Gelfand, I. M., Fomin, S. V., Silverman, R. A. (tr), 2000, *Calculus of Variations* (Dover Publications, Mineola, NY, 2000).

Giaquinta, M., Modica, G., Souchek, J., 1998, *Cartesian Coordinates in the Calculus of Variations* Volumes I and II (Springer-Verlag, New York, 1998).

Giaquinta, M., Hildebrandt, S., 1996, *Calculus of Variations* Volumes I and II (Springer-Verlag, New York, 1996).

Gradshteyn, I. S. and Ryzhik, I. M., 1965, *Table of Integrals, Series, and Products* (Academic Press, New York, 1965).

Heitler, W., 1954, *The Quantum Theory of Radiation* (Claendon Press, Oxford, UK, 1954).

Huang, Kerson, 1992, *Quarks, Leptons & Gauge Fields 2nd Edition* (World Scientific Publishing Company, Singapore, 1992).

Jost, J., Li-Jost, X., 1998, *Calculus of Variations* (Cambridge University Press, New York, 1998).

Kaku, Michio, 1993, *Quantum Field Theory*, (Oxford University Press, New York, 1993).

Kirk, G. S. and Raven, J. E., 1962, *The Presocratic Philosophers* (Cambridge University Press, New York, 1962).

Landau, L. D. and Lifshitz, E. M., 1987, *Fluid Mechanics 2nd Edition*, (Pergamon Press, Elmsford, NY, 1987).

Misner, C. W., Thorne, K. S., and Wheeler, J. A., 1973, *Gravitation* (W. H. Freeman, New York, 1973).

Rescher, N., 1967, *The Philosophy of Leibniz* (Prentice-Hall, Englewood Cliffs, NJ, 1967).

Rieffel, Eleanor and Polak, Wolfgang, 2014, *Quantum Computing* (MIT Press, Cambridge, MA, 2014).

Riesz, Frigyes and Sz.-Nagy, Béla, 1990, *Functional Analysis* (Dover Publications, New York, 1990).

Sagan, H., 1993, *Introduction to the Calculus of Variations* (Dover Publications, Mineola, NY, 1993).

Sakurai, J. J., 1964, *Invariance Principles and Elementary Particles* (Princeton University Press, Princeton, NJ, 1964).

Streater, R. F. and Wightman, A. S., 2000, *PCT, Spin, Statistics, and All That* (Princeton University Press, Princeton, NJ 2000).

Weinberg, S., 1972, *Gravitation and Cosmology* (John Wiley and Sons, New York, 1972).

Weinberg, S., 1995, *The Quantum Theory of Fields Volume I* (Cambridge University Press, New York, 1995).

INDEX

About the Author

Stephen Blaha is a well-known Physicist and Man of Letters with interests in Science, Society and civilization, the Arts, and Technology. He had an Alfred P. Sloan Foundation scholarship in college. He received his Ph.D. in Physics from Rockefeller University. He has served on the faculties of several major universities. He was also a Member of the Technical Staff at Bell Laboratories, a manager at the Boston Globe Newspaper, a Director at Wang Laboratories, and President of Blaha Software Inc. and of Janus Associates Inc. (NH).

Among other achievements he was a co-discoverer of the "r potential" for heavy quark binding developing the first (and still the only demonstrable) non-Aeolian gauge theory with an "r" potential; first suggested the existence of topological structures in superfluid He-3; first proposed Yang-Mills theories would appear in condensed matter phenomena with non-scalar order parameters; first developed a grammar-based formalism for quantum computers and applied it to elementary particle theories; first developed a new form of quantum field theory without divergences (thus solving a major 60 year old problem that enabled a unified theory of the Standard Model and Quantum Gravity without divergences to be developed); first developed a formulation of complex General Relativity based on analytic continuation from real space-time; first developed a generalized non-homogeneous Robertson-Walker metric that enabled a quantum theory of the Big Bang to be developed without singularities at t = 0; first generalized Cauchy's theorem and Gauss' theorem to complex, curved multi-dimensional spaces; received Honorable Mention in the Gravity Research Foundation Essay Competition in 1978; first developed a physically acceptable theory of faster-than-light particles; first derived a composition of extremums method in the Calculus of Variations; first quantitatively suggested that inflationary periods in the history of the universe were not needed; first proved Gödel's Theorem implies Nature must be quantum; provided a new alternative to the Higgs Mechanism, and Higgs particles, to generate masses; first showed how to resolve logical paradoxes including Gödel's Undecidability Theorem by developing Operator Logic and Quantum Operator Logic; first developed a quantitative harmonic oscillator-like model of the life cycle, and interactions, of civilizations; first showed how equations describing superorganisms also apply to civilizations. A recent book shows his theory applies successfully to the past 14 years of history and to *new* archaeological data on Andean and Mayan civilizations as well as Early Anatolian and Egyptian civilizations.

He first developed an axiomatic derivation of the form of The Standard Model from geometry – space-time properties – The Unified SuperStandard Model. It unifies all the known forces of Nature. It also has a Dark Matter sector that includes a Dark ElectroWeak sector with Dark doublets and Dark gauge interactions. It uses quantum coordinates to remove infinities that crop up in most

interacting quantum field theories and additionally to remove the infinities that appear in the Big Bang and generate inflationary growth of the universe. It shows gravity has a MOND-like form without sacrificing Newton's Laws. It relates the interactions of the MOND-like sector of gravity with the r-potential of Quark Confinement. The axioms of the theory lead to the question of their origin. We suggest in the preceding edition of this book it can be attributed to an entity with God-like properties. We explore these properties in "God Theory" and show they predict that the Cosmos exists forever although individual universes (or incarnations of our universe) "come and go." Several other important results emerge from God Theory such a functionally triune God. The Unified SuperStandard Theory has many other important parts described in the Current Edition of *The Unified SuperStandard Theory* and expanded in subsequent volumes.

Blaha has had a major impact on a succession of elementary particle theories: his Ph.D. thesis (1970), and papers, showed that quantum field theory calculations to all orders in ladder approximations could not give scaling deep inelastic electron-nucleon scattering. He later showed the eigenvalue equation for the fine structure constant α in Johnson-Baker-Willey QED had a zero at $\alpha = 1$ not 1/137 by solving the Schwinger-Dyson equations to all orders in an approximation that agreed with exact results to 4th order in α thus ending interest in this theory. In 1979 at Prof. Ken Johnson's (MIT) suggestion he calculated the proton-neutron mass difference in the MIT bag model and found the result had the wrong sign reducing interest in the bag model. These results all appear in Physical Review papers. In the 2000's he repeatedly pointed out the shortcomings of SuperString theory and showed that The Standard Model's form could be derived from space-time geometry by an extension of Lorentz transformations to faster than light transformations. This deeper space-time basis greatly increases the possibility that it is part of THE fundamental theory. Recently, Blaha showed that the Weak interactions differed significantly from the Strong, electromagnetic and gravitation interactions in important respects while these interactions had similar features, and suggested that ElectroWeak theory, which is essentially a glued union of the Weak interactions and Electromagnetism, possibly modulo unknown Higgs particle features, be replaced by a unified theory of the other interactions combined with a stand-alone Weak interaction theory. Blaha also showed that, if Charmonium calculations are taken seriously, the Strong interaction coupling constant is only a factor of five larger than the electromagnetic coupling constant, and thus Strong interaction perturbation theory would make sense and yield physically meaningful results.

In graduate school (1965-71) he wrote substantial papers in elementary particles and group theory: The Inelastic E- P Structure Functions in a Gluon Model. Phys. Lett. B40:501-502,1972; Deep-Inelastic E-P Structure Functions In A Ladder Model With Spin 1/2 Nucleons, Phys.Rev. D3:510-523,1971; Continuum Contributions To The Pion Radius, Phys. Rev. 178:2167-2169,1969; Character Analysis of U(N) and SU(N), J. Math. Phys. 10, 2156 (1969); and The Calculation of the Irreducible Characters of the Symmetric Group in Terms of the

Compound Characters, (Published as Blaha's Lemma in D. E. Knuth's book: *The Art of Computer Programming Vols. 1 – 4*).

In the early 1980's Blaha was also a pioneer in the development of UNIX for financial, scientific and Internet applications: benchmarked UNIX versions showing that block size was critical for UNIX performance, developing financial modeling software, starting database benchmarking comparison studies, developing Internet-like UNIX networking (1982) and developing a hybrid shell programming technique (1982) that was a precursor to the PERL programming language. He was also the manager of the AT&T ten-year future products development database. His work helped lead to commercial UNIX on computers such as Sun Micros, IBM AIX minis, and Apple computers.

In the 1980's he pioneered the development of PC Desktop Publishing on laser printers and was nominated for three "Awards for Technical Excellence" in 1987 by PC Magazine for PC software products that he designed and developed.

Recently he has developed a theory of Megaverses – actual universes of which our universe is one – with quantum particle-like properties based on the Wheeler-DeWitt equation of Quantum Gravity. He has developed a theory of a baryonic force, which had been conjectured many years ago, and estimated the strength of the force based on discrepancies in measurements of the gravitational constant G. This force, operative in D-dimensional space, can be used to escape from our universe in "uniships" which are the equivalent of the faster-than-light starships proposed in the author's earlier books. Thus travel to other universes, as well as to other stars is possible.

Blaha also considered the complexified Wheeler-DeWitt equation and showed that its limitation to real-valued coordinates and metrics generated a Cosmological Constant in the Einstein equations.

The author has also recently written a series of books on the serious problems of the United States and their solution as well as a book on the decline of Mankind that will follow from current social and genetic trends in Mankind.

In the past twenty years Dr. Blaha has written over 80 books on a wide range of topics. Some recent major works are: *From Asynchronous Logic to The Standard Model to Superflight to the Stars, All the Universe!, SuperCivilizations: Civilizations as Superorganisms, America's Future: an Islamic Surge, ISIS, al Qaeda, World Epidemics, Ukraine, Russia-China Pact, US Leadership Crisis, The Rises and Falls of Man – Destiny – 3000 AD: New Support for a Superorganism MACRO-THEORY of CIVILIZATIONS From CURRENT WORLD TRENDS and NEW Peruvian, Pre-Mayan, Mayan, Anatolian, and Early Egyptian Data, with a Projection to 3000 AD,* and *Mankind in Decline: Genetic Disasters, Human-Animal Hybrids, Overpopulation, Pollution, Global Warming, Food and Water Shortages, Desertification, Poverty, Rising Violence, Genocide, Epidemics, Wars, Leadership Failure.*

He has taught approximately 4,000 students in undergraduate, graduate, and postgraduate corporate education courses primarily in major universities, and large companies and government agencies.

Recently he developed a quantum theory, The Unified SuperStandard Theory (UST), which describes elementary particles in detail without the difficulties of conventional quantum field theory. He found that the internal symmetries of this theory could be exactly derived from an octonion theory called QUeST. He further found that another octonion theory (UTMOST) describes the Megaverse. It can hold QUeST universes such as our own universe. It has an internal symmetry structure which is a superset of the QUeST internal symmetries.